RNA EDITING
The alteration of protein coding sequences of RNA

RNA EDITING
The alteration of protein coding sequences of RNA

Editor:
ROB BENNE
E.C. Slater Institute, Academic Medical Centre,
Meibergdreef 15, 1105 AZ,
Amsterdam, The Netherlands

ELLIS HORWOOD
NEW YORK LONDON TORONTO SYDNEY TOKYO SINGAPORE

First published in 1993 by
ELLIS HORWOOD LIMITED
Market Cross House, Cooper Street,
Chichester, West Sussex, PO19 1EB, England

A division of
Simon & Schuster International Group
A Paramount Communications Company

Printed and bound in Great Britain by
Hartnolls Ltd, Bodmin, Cornwall

British Library Cataloguing in Publication Data

A catalogue record for this book is available from the British Library

ISBN 0–13–782558–7

Library of Congress Cataloging-in-Publication Data

Available from the publisher

Table of contents

Acknowledgements

First and foremost, I wish to thank Els Vlugt-van Daalen for her skillful and patient handling of countless versions of the chapters. Without her, there most certainly would not have been a book on RNA editing. Next, I would like to thank the authors of the chapters for the time they set aside in their busy academic and scientific lives to produce their excellent contributions and the patience with which they responded to my queries for manuscripts, figures, comments on alterations, etc. I also gratefully acknowledge the past and present members of the trypanosome group for their support, both moral and scientific and I apologize to them and all the other colleagues of the E.C. Slater Institute for any inconvenience that was caused by my spending so much time in assembling this book. Special thanks are due to Duco Zonneveld for designing the cover and his expert help in redrawing some of the figures, to Paul Sloof for critically reading a number of the chapters, and to Cars Gravemeijer for photographically reproducing most of the figures in the correct size. Last, but certainly not least, I am greatly indebted to Wendy van Noppen for her reading of all the manuscripts and her invaluable help in the editing of the book on editing.

Abbreviations

aa	amino acids:
Ala, A	alanine
Arg, R	arginine
Asn, N	asparagine
Asp, D	aspartic acid
Cys, C	cysteine
Gln, Q	glutamine
Glu, E	glutamic acid
Gly, G	glycine
His, H	histidine
Ile, I	isoleucine
Leu, L	leucine
Lys, K	lysine
Met, M	methionine
Phe, F	phenylalanine
Pro, P	proline
Ser, S	serine
Thr, T	threonine
Trp, W	tryptophane
Tyr, Y	tyrosine
Val, V	valine

12S, 9S	large, small rRNA, respectively, of trypanosomes
A	adenine, adenosine
A6	ATPase subunit 6
ApoB	Apolipoprotein B
atp	gene encoding ATPase
ATP	Adenosine triphosphate

ATPase, ATP syn	ATP synthase
b	bovine
bp	base pair
C	cytosine, cytidine
cDNA	copy DNA
CNS	central nervous system
CO	cytochrome c oxidase
cob	gene encoding apocytochrome b
cox	(gene encoding) cytochrome c oxidase
cp	chloroplast
CR1-6	maxicircle regions with a cytosine-rich template strand
CSB	conserved sequence block
CTP	cytidine triphosphate
CYb, Cyt b	apocytochrome b
DCS	domain connecting sequences
ds	double stranded
EBS	exon binding site
FGF	fibroblast growth factor
G	guanine, guanosine
G1-6	maxicircle regions encoding G-rich RNAs (= CR1-6)
GluR	glutamate gated channel subunit
gRNA	guide RNA
GTP	guanosine triphosphate
h	human
HBV	hepatitis B virus
HDV	hepatitis delta virus
I	inosine
IBS	intron binding site
ITP	inosine triphosphate
k	kinetoplast
kb	kilo base pair
kDa	kilodalton
L.t.	*Leishmania tarentolae*
LDL	low density lipoprotein
LPMV	La-Piedad-Michoacan-Mexico-Virus
LSU	large ribosomal subunit
mRNA	messenger RNA
mt	mitochondrial
MURF	maxicircle unidentified reading frame
MV	measles virus
nad, ndh	gene encoding NAD, NDH
NC	nucleocapsid
ND, NAD, NDH	NADH dehydrogenase
NDV	Newcastle disease virus
NP	nucleoprotein
nt	nucleotide

ORF	open reading frame
P. polycephalum	*Physarum polycephalum*
PCR	polymerase chain reaction
PER	pre-edited region
PIV	parainfluenza virus
PL	proteolipid
ps6	photosystem 6
rpl	protein of the large ribosomal subunit
rps	protein of the small ribosomal subunit
rRNA	ribosomal RNA
RT	reverse transcriptase
SEN	Sendai virus
ss	single stranded
SSU	small ribosomal subunit
SV	simian virus
T	thymine, thymidine
T.b.	*Trypanosoma brucei*
TM	transmembrane
tRNA	transfer RNA
TTP	thymidine triphosphate
U	uracil, uridine
URF	unidentified reading frame
UTP	uridine triphosphate

1

RNA EDITING, AN OVERVIEW

Rob Benne
E.C. Slater Institute, University of Amsterdam, Academic Medical Centre,
Meibergdreef 15, 1105 AZ Amsterdam, The Netherlands

1.1 INTRODUCTION AND PURPOSE OF THIS BOOK

Proteins play a pivotal role in determining the function and structure of a living cell. Not only are proteins an integral part of macromolecular elements, such as cytoplasmic and nuclear skeletons, membranes etc. that give shape to the cell and its organelles, proteins are also major constituents of the enzymatic machineries that operate in the biosynthesis of essential macromolecules. Moreover, proteins enable a cell to extract from the available nutrients the energy required to sustain life. It therefore goes without saying that cellular existence depends on the ability to produce these proteins at the appropriate time and in the correct amount without mistakes in their amino acid sequence, and to pass on this ability to the daughter cells that arise from cell division. A vast amount of research done over the last five decades has revealed the principal elements of how this is achieved. The key series of processes are contained within the "Central Dogma" of molecular biology. The dogma states that the information for the amino acid sequence for the proteins resides in the nucleotide sequence of the genes of the cellular DNA and that this information is delivered to the protein synthetic machinery through an RNA intermediate, which is copied from one of the two DNA strands. In short, the Central Dogma states that 'DNA makes RNA makes protein'.

In the early days most of the work on gene expression was done in bacterial systems. Each amino acid in a protein could be directly traced back to its corresponding codon in the DNA and successive amino acids were specified by a

organism	type of RNA editing	example of edited sequence	nr of edited nucleotides	cellular location	mechanism, chapter
trypano- somes	U insertion/ deletion	Ser Gly Glu Glu Lys AuGuuuCGuuGuAGAuuuuAuuAuuuuuuuuAuuA Met Phe Arg Cys Arg Phe Leu Leu Phe Phe Leu Leu	3905, T. brucei	mitochondrion	transesterification directed by gRNAs, 2, 3
Physarum	mixed nucleotide insertion	Val Val Leu STOP GUCcGUGCUUUAAAUACcUUAGUCAAAACCCcU Val Arg Ala Leu Asn Thr Leu Val Lys Thr Pro	257	mitochondrion	?, 4
paramyxo viruses	G insertion	Phe Lys Arg Gly Arg Asp Thr UUUAAGAGAGGGGggCAGGGAUACCG Phe Lys Arg Gly Ala Gly Ile Pro	2, SV5 1, SEN	cytoplasm	co-transcriptional RNA polymerase stuttering, 5
mammals	C → U conversion	Gln Thr Tyr Met Ile Gln Phe Asp CAGACAUAUAUGAUA C AAUUUGAU Gln Thr Tyr Met Ile STOP	2	nucleus	C-deamination, 6
mammals	A → G (or I) conversion	Phe Met Gln Gln UUUAUGC A GCAA Phe Met Arg Gln	1, GluR 133, MVmp	nucleus(?)	A-deamination, 8
higher plants	C → U conversion	Val Arg Leu Phe Pro Phe Val GUU C GGUUAUCC C AUUUGUC Val Trp Leu Phe Leu Phe Val	134, wheat	mitochondrion, chloroplast	?, 7
higher plants	U → C	Asp Ile Phe Phe GAUAUC U UUUUC Asp Ile Leu Phe	1, wheat	mitochondrion	?, 7
hepatitis delta virus [a]	U → C conversion	Asp Ile Phe Phe GAUAUC c UUUUC Asp Ile Leu Phe	1, HDV	nucleus	U-amination, 1, 7

reading frame of successive codons. There was no doubt that one gene encodes only one protein. The discovery of splicing in eukaryotes complicated this relatively simple picture. The colinearity principle and the 'one gene one protein' rule had to be abandoned. Adjacent amino acids in a protein can be encoded by triplets "miles apart" in the DNA, even codons internally split by an intron are not unusual, and alternatively spliced RNAs appeared on the scene (one gene, more than one protein). It was further realized that apart from the triplet code (the 'genetic' code), the genome also contains other codes, such as signals for splicing, ribosome binding to RNA, protein routing etc. What still appeared to be true, though, is the fact that the identity of an amino acid of a protein can be directly derived from the nucleotide sequence of DNA via the application of the genetic code, although some caution is required in the rare cases in which the translating ribosomes switch from one reading frame to another.

The discovery in recent years in a number of genetic systems of RNAs that are edited, and therefore contain nucleotides that are different from those encoded by the genome, has eliminated even this last certainty. As it turns out, on top of the genetic code other signals are superimposed which ensure that an RNA is edited. The identification of these signals and the elucidation of the different mechanisms by which RNAs can be edited is therefore essential for a complete understanding of the way genes are expressed. This book aims at describing in a comprehensive fashion the state of the art of the research on RNA editing in a number of organisms. Each chapter contains a brief description of the genetic system in question and, although the editing processes can be very different, the same questions are asked in each case: (i) how is an editing site selected? (ii) what is the biochemical mechanism of editing? and (iii) what could be its evolutionary origin?

Fig. 1 —— RNA editing alters the protein coding sequence of transcripts.
The figure presents a list of RNA segments from various organisms, which are edited in different ways. The nucleotides resulting from editing are in bold lower case lettering, for example 24 Us of the *Trypanosoma brucei* CYb sequence shown are the result of an insertional editing process. Unedited nucleotides are in capital letters. The genomically encoded amino acids are given above the RNAs, the RNA-based amino acids below. The triplet sequences that are translated into proteins are shaded. In most cases they are encoded by edited RNAs. However, for the paramyxovirus P/V RNAs and the mammalian apolipoprotein B and glutamate receptor RNAs both the edited and the unedited RNAs are used for protein synthesis. Abbreviations: cox, cytochrome c oxidase; CYb, apocytochrome b; ATPase, ATP synthase; ApoB, apolipoprotein B; GluR, glutamate gated receptor channel subunit; HDV, hepatitis delta virus; SEN, Sendai virus; MVmp, measles virus matrix protein. The following examples of edited RNA segments are shown: trypanosomes, CYb RNA from *T. brucei*; *Physarum polycephalum*, ATPase αRNA; paramyxoviruses, SV5 P/V RNA; mammals, ApoB 100 RNA (C-to-U conversion) and GluR RNA (A-to-G/I conversion); higher plants, cox2 RNA (C-to-U conversion) and cox3 RNA (U-to-C conversion). The number of edited nucleotides is updated to July 1992.
[a]The U to C conversion in hepatitis delta virus (HDV) occurs in the genomic RNA strand of the virus. The affected viral protein (the delta antigen), however, is synthesized from the <u>anti</u>-genomic strand, which contains the corresponding A-to-G change. Also in this case both the edited and unedited RNA are used for protein synthesis.

1.2   DIFFERENT FORMS OF RNA EDITING EXIST IN DIFFERENT GENETIC SYSTEMS

The term 'RNA editing' was first used to explain the presence of four uridines in the mitochondrial (mt) transcript for cytochrome c oxidase (cox) subunit 2 of trypanosomes that are not encoded in the DNA (Benne *et al.*, 1986). It was coined to illustrate that the alterations of the RNA sequence (i) occur in the protein-coding region and (ii) are most likely the result of a posttranscriptional event. As described in Chapters 2 and 3, numerous examples of similarly altered trypanosome mt transcripts have been found. All the available evidence suggests that in these organelles RNAs are edited by sometimes massive posttranscriptional U-insertion and U-deletion processes (see Fig. 1).

The term RNA editing has also been used for other RNA sequence-alteration processes, examples of which are listed in Fig. 1. Depending on the nature of the alteration they can be roughly subdivided into processes that involve insertion/deletion of nucleotides and those that entail nucleotide conversion. The insertion of Cs and other nucleotides into mt RNAs of *Physarum polycephalum* (Mahendran *et al.*, 1991; Chapter 4), insertion of Gs into paramyxoviral P-RNAs (Thomas *et al.*, 1988; Chapter 5), and poly[A] addition to the 3' end of mammalian mt transcripts (Ojala *et al.*, 1981; not included) belong to the first category. The C to U conversion of the mammalian apolipoprotein B transcript (Chen *et al.*, 1987; Powell *et al.*, 1987; Chapter 6), the pyrimidine interconversions in plant organellar RNAs (Covello and Gray, 1989; Gualberto *et al.*, 1989; Hiesel *et al.*, 1989; Hoch *et al.*, 1991; Chapter 7), the U to C conversion of the genomic RNA of human hepatitis delta virus (HDV) (Zheng *et al.*, 1992; Taylor *et al.*, 1992), and the A to G (or I) change in human glutamate receptor RNA (Sommer *et al.*, 1991; Chapter 8) to the second.

In virtually all cases the sequence of the edited RNAs has exclusively been inferred from cDNA analysis and only for the edited apolipoprotein B transcript has the identity of the edited U been confirmed by nucleotide analysis of the RNA (Hodges *et al.*, 1991; see Chapter 6). Moreover, the ApoB RNA and plant mt atp9 RNA (Begu *et al.*, 1990; see Chapter 7) provide the only examples for which the effects of editing have been verified at the protein level. The theoretical possibility that edited nucleotides in other systems are not Us, Cs , etc., but unusual, modified nucleotides that are only read as Us or Cs by reverse transcriptase, has therefore not rigorously been excluded. Given the problems unusual nucleotides might pose to the protein synthetic machinery the general feeling is, however, that in most editing processes conventional nucleotides are created. The only exception to this could be the editing of the mammalian glutamate gated channel subunit (GluR) RNA, for which the primary modified nucleotide is proposed to be an inosine. This inosine shows up as a G in the cDNA analysis that was performed (see Table 1 and Chapter 8). However, whether inosine-containing mRNAs indeed exist and are used for protein synthesis remains to be verified.

In all cases of RNA editing, experiments to search for other, as yet unnoticed versions of the genes in question that could encode the altered transcripts were unsuccessful. The RNAs therefore seem to be derived from bona fide RNA editing events. Not all processes, however, appear to occur posttranscriptionally. For example, the G-insertion process of paramyxoviral P-RNAs is proposed to be the result of a 'stuttering' RNA polymerase (Chapter 5). It should be stated, however, that the current evidence for the posttranscriptional nature of most RNA editing processes is indirect, based largely on the occurrence and characteristics of partially altered transcripts (Chapters 2, 3, 4 and 7), or lacking altogether (Chapter 8). Only the C to U conversion of apolipoprotein B RNA has been unambiguously proved to be a posttranscriptional process with the aid of an *in vitro* editing system (Driscoll *et al.*, 1989; Chapter 6), and for the time being it does not appear very useful, although maybe semantically correct, to reserve the term RNA editing exclusively for posttranscriptional processes. For the purpose of this book, therefore, RNA editing is defined as any process that produces alterations in protein-coding sequences of an RNA. As illustrated in Fig. 1, the majority of the editing events described so far takes place inside mitochondria, but editing is also found in the nucleus, the chloroplast and the cytoplasm.

In time, the definition of RNA editing may be subject to even further changes, since the growing list of edited RNAs provides ample evidence that the same processes that alter nucleotides of protein-coding regions of RNA also operate elsewhere. U-insertion/deletion processes change leader and trailer sequences of trypanosome mt mRNAs (Chapters 2 and 3). Mammalian tRNAs (Diamond *et al.*, 1990) and intron sequences of plant mt precursor RNAs (Chapter 7) contain 'edited' pyrimidine residues. Last but not least, ribosomal RNAs and tRNAs in *P. polycephalum* contain multiple inserted nucleotides (Chapter 4). Although the functional importance of these alterations remains to be assessed, editing processes affect more than the coding potential of an RNA and are possibly also used to alter other genomic signals. In that case the term RNA editing could be applicable to virtually any co- or posttranscriptional alteration of the identity of an RNA nucleotide.

1.3   SITE SELECTION IN RNA EDITING

What then are the signals that designate a certain nucleotide sequence for alteration? It is, *a priori*, not illogical to assume that primary sequences and/or secondary structure motifs around editing sites play a role. As shown in Chapter 6, the site of the C to U conversion of apolipoprotein B RNA is indeed selected with the aid of a short sequence motif a few nucleotides downstream. Although the presence of the editing activity in cells that do not produce apolipoprotein B suggests that other mammalian mRNAs are also edited by this particular mechanism, so far the apolipoprotein B RNA provides the only example(s) and the sequence motif has not yet been spotted anywhere else. The consensus opinion, however, is that the machinery for this particular way of editing has not evolved for the alteration of just one nucleotide of RNA and that other examples will be found.

Other examples of limited editing are provided by the G-insertions in paramyxovirus P RNAs (Chapter 5), the U to C conversion in the genomic RNA of HDV and the A to G(or I) change in the human glutamate receptor transcript (Chapter 8). Although the mechanisms of these editing processes appear to be completely different from each other and, except for the HDV editing from that of the apolipoprotein B RNA, site selection could in principle be mediated by a relatively simple primary or secondary sequence motif. Such motifs, however, have not yet been identified, although preliminary evidence suggests that editing of the HDV RNA requires basepairing of a small region around the editing site to another part of the RNA (Taylor *et al.*, 1992). For A to G(or I) changes, this relatively simple picture is complicated by the observation that multiple A to G changes, at first sight similar to the one in GluR cDNA, occur at random positions in the genome of defective measles virus (MV) in human brain infections (Cattaneo *et al.*, 1988; see Fig. 1). If these G to A changes are indeed created by the same enzymatic machinery that edits GluR RNA, the large difference in specificity of site selection must be explained.

In other genetic systems RNAs are much more frequently edited. For example, so far 134 editing sites have been identified in wheat mt RNAs (Chapter 7) and in trypanosome mtRNAs thousands of Us are inserted or deleted (3905 in *T. brucei*!) at countless sites (Chapters 2 and 3). Also in *P. polycephalum* RNAs (Chapter 4) numerous editing sites have been found. Inspection of the sequences around these sites in all of these organisms has repeatedly failed to produce any obvious common motif that could be involved in site selection. In trypanosomes an essential part of the mystery of site selection has been solved by the discovery of small so-called guide (g)RNAs (Blum *et al.*, 1990; Chapters 2 and 3). Each gRNA provides the genetic information for editing of a small segment of RNA on which a limited number of editing sites is located. The editing sites are selected by virtue of the fact that the gRNAs can basepair with the target pre-edited RNA 3' of the region to be edited. The large number of editing sites and the small size of the gRNAs implies that many gRNAs are required. For example, in *Leishmania tarentolae* 26 different gRNAs have already been identified (Chapter 3) and in *T. brucei* the actual number will turn out to be much higher.

Given the comparably high number of editing sites, it has been speculated that gRNAs or some other form of antisense nucleic acid are also involved in the editing in *P. polycephalum* (Chapter 4) and plant mitochondria (Chapter 7). The experimental evidence that would support such a proposal is still lacking, however.

1.4   MECHANISMS OF RNA EDITING

The availability of an RNA editing system *in vitro* has made the C to U conversion reaction of mammalian apolipoprotein B RNA the best characterized RNA editing process. The available evidence suggests that the editing activity is a protein(s) which catalyses a site-specific C-deamination reaction (Chapter 6). The protein(s) involved have not yet been purified and characterized, but this is actively being pursued in a number of laboratories. U to C editing of the HDV RNA is stimulated *in vitro* by glutamine and inhibited by a glutamine analogue, suggesting that the editing enzyme is like CTP-synthetase. This enzyme converts UTP to CTP in prokaryotic and eukaryotic cells. As stated above, the editing of paramyxovirus P RNA is brought about, most likely, by some sort of stuttering mechanism during transcription (Chapter 5). This form of editing can also be studied *in vitro* and further characterisation of the proposed components involved appears relatively straightforward.

In all other genetic systems which display RNA editing, the lack of an *in vitro* system has thus far hampered rapid progress and most of the current ideas have been derived from the analysis of the edited RNAs and, in trypanosomes, of the gRNAs. Particularly the notion that gRNAs are complementary to edited RNA segments has been elementary in the development of the concept that they provide the edited information, albeit not as conventional templates since the complementarity also includes G:U basepairing. Knowledge of the precise mechanism of action of the gRNAs is still scarce, but recent breakthroughs (Chapters 2 and 3) have been the realization that gRNAs have a 3' terminal oligo [U] extension (Blum and Simpson, 1990) and the identification of so-called chimeric molecules, in which via this U-tail a gRNA is covalently linked to an editing site of the target pre-edited RNA (Blum *et al.*, 1991). The existence of these molecules has led to the proposal that the inserted Us in trypanosome mtRNAs are provided by the gRNAs and that the U-tail is also instrumental in the deletion process. Whether this is done via a series of concerted RNA-mediated transesterification reactions, in analogy to splicing (Blum *et al.*, 1991, see also Cech, 1991), or via the involvement of 'cut and paste' enzymes is still a matter of debate. This can only be settled with a functional system *in vitro*. Interestingly, preliminary evidence suggests that gRNAs in mt lysates are present in large RNP complexes. The identification and characterisation of proteins that are part of these complexes and of other proteins that may play a role in RNA editing in trypanosomes has been initiated in several laboratories (see Chapters 2 and 3).

The study of extensively edited RNAs in trypanosomes indicates that the overall polarity of the editing process is from 3' to 5'. There is some disagreement, though, on the precise sequence of events taking place during the editing directed by a single gRNA. One school of thought is represented by Stuart (Chapter 2) preferring the view that this process in *T. brucei* does not have a strict polarity and that it depends on the precise way in which basepairing between gRNA and pre-edited RNA can take place. The results obtained in *L. tarentolae*, however, have led to a slightly different interpretation (Chapter 3), in which the editing process has a strict 3' to 5' polarity, operating at one site at a time. For a more detailed discussion of the current concepts and models of trypanosome editing, I refer to Chapters 2 and 3. Again, such a dispute can only be settled with an *in vitro* system in which it can be established which molecules are true intermediates of the editing process and which are dead-end side products. The recent development *in vitro* of systems capable of stimulating the formation of chimeric molecules (Harris and Hajduk, 1992; Koslowsky *et al.*, 1992) is a guarantee that further insight in the mechanism of RNA editing in trypanosomes will be forthcoming.

In *P. polycephalum* (Chapter 4), no preferred polarity of the editing process has been observed given the fact that edited nucleotides can be found at either end of partially edited RNAs. In plant mitochondria there appears to be no directionality at all, since edited nucleotides are present at random positions in a partially edited RNA (Chapter 7). As stated in the previous section, the trypanosome results have been an incentive to search for gRNA-like molecules in *P. polycephalum* and plant mitochondria. Although it is attractive to speculate on the possible involvement of such molecules in these systems, efforts to prove their existence have not been successful so far and no definitive clues are available as to the mechanism of these two types of RNA editing (see Chapters 4 and 7).

As indicated above, even more speculation is required with respect to the mechanism of the A to G conversion of the human glutamate receptor RNA. For reasons discussed in Chapters 6 and 8, the most likely (or least unlikely) possibility would be the involvement of a double-stranded RNA-specific A-deaminase. Double stranded RNA A-deaminase is also thought to be responsible for the biased A$\rightarrow$G hypermutations in the genomes of MV and other viruses, see Chapter 8. In theory, they could also be produced by U-to-C changes in the antigenome e.g. when duplexes have formed during replication. The apparent randomness of the alterations, however, seems to argue in favour of a role for ds RNA A-deaminase, since the U-to-C change in HDV is a highly specific single site conversion. Further work, preferably *in vitro* should settle this issue.

1.5   THE FUNCTION AND EVOLUTION OF RNA EDITING

There are a number of lessons to be learned from the RNA editing processes described in this book. Firstly, there is the realization that the genomic nucleotide sequences contain signals that result in the alteration of the identity of RNA nucleotides. These editing signals may be sequences to which a guide RNA can basepair (Chapter 2 and 3), sequences which allow 'backsliding' of the nascent RNA on the template (Chapter 5), or sequences that help an enzyme to select a certain C for conversion into U (or vice versa, Chapters 6 and 7) or an A into a G (Chapter 8). As long as we do not know these signals (and we do not in most cases) and the way in which the nucleotide order of the RNA will be changed, some areas of a genomic sequence will be cryptic, i.e. they do not encode the amino acid sequence we think they do. Nevertheless, the central dogma 'DNA makes RNA makes protein' still holds, since all the editing signals are encoded in the genome.

At least as important as the 'how' is the 'why' of RNA editing: why do some selected parts of the genetic message have to be altered at the level of the RNA? There is no easy answer to this question; we can only look at what the consequences are of the different editing processes and do some creative speculating. The effects of editing on the production of proteins in the different systems are summarized in Table 1. The result of the paramyxoviral G insertion

Table 1 — Editing affects the production of proteins

Type of editing (organism)	Effect on protein production
U-insertion/deletion (trypanosomes)	mRNAs are made translatable: - removal of frameshifts - creation of start/stop codons - creation of complete coding sequences
Mixed nucleotide insertion (*Physarum*)	mRNAs are made translatable: - removal of frameshifts
G-insertion (paramyxoviruses)	Two or three proteins from one gene: - removal/creation of frameshifts in some of the RNAs
C to U conversion (mammals)	Two proteins from one gene: - creation of a stop codon in some of the RNAs
A to G (or I) conversion (mammals)	Two proteins from one gene: - one codon change in some of the RNAs
Pyrimidine interconversion (higher plants)	mRNAs are made translatable: - creation of start/stop codons - multiple codon changes
U to C conversion (HDV)	Two proteins from one gene: - elimination of a stop codon in some of the RNAs
Poly(A) addition (mammals)	mRNAs are made translatable - creation of stop codons

and the HDV U to C change is that more than one protein is encoded by one gene, since <u>both</u> the edited and the unedited RNAs are used as mRNA. Keeping their genome as small as possible is particularly important for viruses if they want to survive and space-saving processes other than RNA editing are routinely employed. Also the creation of UAA stopcodons by polyadenylation of mammalian mt RNAs can be considered as a space-saving process. The gene organisation of mammalian mtDNA is highly compact with very few, if any, nucleotides separating the coding regions of adjacent genes. This drive for compactness has even resulted in the elimination of some of the As of the UAA stopcodons and complete stopcodons have to be created by polyadenylation of the transcripts. Space is clearly not a problem for mammalian nuclear DNA, so the rationale is not so obvious for the C to U conversion of apolipoprotein B RNA (Chapter 6) and the A to G change of the glutamate gated channel subunit (Chapter 8). Also there, however, we see one gene encoding two proteins. In mammalian genomes, gene duplication followed by separate evolution of the two copies would be a more obvious way of producing closely related proteins in regulatable amounts. RNA editing, however, does provide the opportunity to introduce highly specific, local changes into only some of the molecules. This results in truncation of the molecule, as in apolipoprotein B, or in alteration of just one amino acid, as in the glutamate receptor. It could be reasoned that somehow this would be more difficult to achieve via gene duplication, since independently accumulating mutations would make it harder to keep the remainder of the two sequences identical.

RNA editing in trypanosome mitochondria (Chapters 2 and 3) and in plant mitochondria (Chapter 7) could also, at least in theory, offer the possibility of producing multiple proteins from one gene. Direct sequencing of proteins (which could prove or disprove this point) has only just begun, however. In these genetic systems [as in *Physarum* mitochondria (Chapter 4)], RNA editing is more likely meant to provide an extra level of regulation of gene expression. As outlined in Table 1, without editing there is no protein-coding sequence that makes any sense and often the start and stop signals for translation are missing. It is not easy to understand, though, why such an extra level of regulation would be required.

The 'raison d'être' of the editing processes is intimately related to their respective evolutionary backgrounds, the crucial question being whether editing is 'old' or 'young' in evolutionary terms. The same question has been asked for the other processes of gene expression and, given their overall similarity in widely different organisms, the answer is usually that they are old processes, which arose before these species diverged billions of years ago. For example, translation is carried out in bacterial and mammalian cells in essentially the same way. Things are not as clear-cut for the RNA editing processes, because mechanistically very different forms appear to exist. A stuttering polymerase, a C-deaminase and a gRNA based U-insertion/deletion machinery operate on the basis of different mechanistic principles and these processes most likely do not have a common evolutionary background. All editing processes, however, somehow involve <u>RNA</u>, and the current notion is that RNA or RNA-like molecules may have been among the earliest macromolecules to evolve on earth (Joyce, 1989). The prevalent opinion therefore is that RNA editing processes may be 'old' in an evolutionary sense, although different forms have arisen (or persisted) in different lineages. For a more extensive discussion of the evolution of each of the editing processes, I refer to the chapters.

REFERENCES

Begu, D., Graves, P.V., Domec, C., Arselin, G., Litvak, S. & Arraya, A. (1990) RNA editing of wheat mitochondrial ATP synthase subunit 9: Direct protein and cDNA sequencing. *Plant Cell* **2** 1283-1290

Benne, R., van den Burg, J., Brakenhoff, J.P.J., Sloof, P., van Boom, J.H. & Tromp, M.C. (1986) Major transcript of the frameshifted coxII gene from trypanosome mitochondria contains four nucleotides that are not encoded in the DNA. *Cell* **46** 819-826.

Blum, B., Bakalara, N. & Simpson, L. (1990) A model for RNA editing in kinetoplast mitochondria: 'guide' RNA molecules transcribed from maxicircle DNA provide the edited information. *Cell* **60** 189-198.

Blum, B. & Simpson, L. (1990) Guide RNAs in kinetoplastid mitochondria have a nonencoded 3' oligo(U) tail involved in recognition of the preedited region. *Cell* **62** 391-397.

Blum, B., Sturm, N.R., Simpson, A.M. & Simpson, L. (1991) Chimeric gRNA-mRNA molecules with oligo(U) tails covalently linked at sites of RNA editing suggest that U addition occurs by transesterification. *Cell* **65** 543-550.

Cattaneo, R., Schmid, P., Eschle, D., Baczko, K., Meallen, V. & Billeter, M.A. (1988) Biased hypermutation and other genetic changes in defective measles virus in human brain infections. *Cell* **55** 255-265.

Cech, T.R. (1991) RNA editing: world's smallest introns? *Cell* **64** 667-669.

Chen, S.H., Habib, G., Yang, C.Y., Gu, Z.W., Lee, B.R., Weng, S.A., Silberman, S.R., Cai, S.J., Deslypere, J.P., Rosseneu, M., Gotto, A.M., Li, W.H. & Chan, L. (1987) apolipoprotein B-48 is the product of a messenger RNA with an organ-specific in-frame stop codon. *Science* **238** 363-366.

Covello, P.S. & Gray, M.W. (1989) RNA editing in plant mitochondria. *Nature* **341** 662-666.

Diamond, A.M., Monteropuerner, Y., Lee, B.J. & Hatfield, D. (1990) Selenocysteine inserting tRNAs are likely generated by tRNA editing. *Nucleic Acids Res.* **18** 6727.

Driscoll, D.M., Wynne, J.K., Wallis, S.C. & Scott, J. (1989) An *in vitro* system for the editing of apolipoprotein B mRNA. *Cell* **58** 519-525.

Gualberto, J.M., Lamattina, L., Bonnard, G., Weil, J.H. & Grienenberger, J.M. (1989) RNA editing in wheat mitochondria results in the conservation of protein sequences. *Nature* **341** 660-666.

Harris, M.E. & Hajduk, S.L. (1992) Kinetoplast RNA editing: *in vitro* formation of cytochrome b gRNA-mRNA chimeras from synthetic substrate RNAs. *Cell* **68** 1091-1099.

Hiesel, R., Wissinger, B., Schuster, W. & Brennicke, A. (1989) RNA editing in plant mitochondria. *Science* **246** 1632-1634.

Hoch, B., Maier, R.M., Appel, K., Igloi, G.L. & Kossel, H., (1991) Editing of a chloroplast mRNA by creation of an initiation codon. *Nature* **353** 178-180.

Hodges, P.E., Navaratnam, N., Greeve, J.C. & Scott, J. (1991) Site specific creation of uridine from cytidine in apolipoprotein B mRNA. *Nucleic Acids. Res.* **19** 1197-1201.

Joyce, G.F. (1989) RNA evolution and the origins of life. *Nature* **338** 217-224.

Koslowsky, D.J., Riley, G.R., Feagin, J.E. & Stuart, K. (1992) In vitro guide RNA/mRNA chimaera formation in *Trypanosoma brucei* RNA editing. *Nature* **356** 807-809.

Mahendran, R., Spottswood, M.R. & Miller, D.L. (1991) RNA editing by cytidine insertion in mitochondria of *Physarum polycephalum. Nature* **349** 434-438.

Ojala, D., Montoya, J. & Attardi, G. (1981) tRNA punctuation model of RNA processing in human mitochondria. *Nature* **290** 470-474.

Powell, L.M., Wallis, S.C., Pease, R.J., Edwards, Y.H., Knott, T.J. & Scott, J. (1987) A novel form of tissue-specific RNA processing produces apolipoprotein-B48 in intestine. *Cell* **50** 831-840.

Sommer, B., Köhler, M., Sprengel, R. & Seeburg, P.H. (1991). RNA editing in brain controls a determinant of ion flow in glutamate-gated channels. *Cell* **67** 11-19.

Taylor, J., Zheng, H.-Q, Fu, T.B., Turner, J., Ruy, W.S., Bayer, M., Netter, H. & Lazinski, D. (1992) 18th EMBO Annual Symposium on RNA, Abstract book 69-70.

Thomas, S.M., Lamb, R.A. & Paterson, R.G. (1988) Two mRNAs that differ by two nontemplated nucleotides encode the amino coterminal proteins P and V 202020of the paramyxovirus SV5. *Cell* **54** 891-902.

Zheng, H., Fu, T.B., Lazinsky, D. & Taylor, J. (1992) Editing on the genomic DNA of human hepatitis delta virus. *J. Virol.* **66** 4693-4697.

2

RNA EDITING IN MITOCHONDRIA OF AFRICAN TRYPANOSOMES

Kenneth Stuart
Seattle Biomedical Research Institute, 4 Nickerson Street, Seattle, Washington
98109-1651, USA

Enigma often precedes new insight, as is the case for the RNA editing in trypanosomes. The kinetoplast (k)DNA of *Trypanosoma brucei* has several enigmatic features that have become comprehensible with the elucidation of RNA editing. *T. brucei* kDNA is composed of about 40 apparently identical 22 kb maxicircles. These are analogous to other mitochondrial DNAs since they encode mitochondrial rRNAs and components of the mitochondrial oxidative phosphorylation system. They are topologically interlocked into a network along with about 10,000 heterogeneous 1 kb minicircles (see Stuart and Feagin, 1992; Englund, 1981; Stuart, 1983a; Benne, 1985; Simpson, 1986; Simpson, 1987; Ryan *et al.*, 1988 for reviews). The minicircles have no known equivalent outside the Kinetoplastida and their function was a mystery until they were found to encode guide (g) RNAs (Sturm and Simpson, 1990b; Bhat *et al.*, 1990; Koslowsky *et al.*, 1990; Sturm and Simpson, 1991) which are complementary to the edited sequences and appear to specify these. The *T. brucei* maxicircles have unexpected sequence features, including frameshifted genes, genes without translational initiation and termination codons, overlapping genes, and several remarkable guanine-versus-cytosine strand-biased sequences whose transcripts are heterogeneous in size and larger than expected from the maxicircle sequence. These unusual features of the *T. brucei* maxicircle reflect editing of its transcripts. This editing revises mRNA coding sequences and in several cases the extent of editing is startling, with the sequence of the mature mRNAs being the result of editing more than of transcription. The pre-edited mRNAs of *T. brucei* resemble first-draft manuscripts. The gRNAs specify the addition and removal of uridines during editing to produce the fully revised mature mRNAs that retain all of the adenines, guanines and cytosines and most of the uridines encoded in the maxicircle gene. Some transcripts undergo minimal revision, such as the addition of four uridines, while others are extensively revised by the addition of hundreds and removal of tens of uridines. Thus, each mature edited mRNA is the product of two or more genes: the maxicircle gene that encodes the pre-edited transcript and the genes that encode the gRNAs that direct the editing.

Regulation of the amounts of mature edited mRNAs that accumulate appears to control the metabolic changes during the life-cycle of *T. brucei*. The life-cycle of African trypanosomes of the brucei group, causal agents of African sleeping sickness in humans and related diseases in animals, alternates between the mammalian hosts and the tsetse fly (*Glossina*) insect vector (Vickerman, 1985). Bloodstream forms initially differentiate into procyclic forms after transmission to the fly midgut and then migrate to the salivary gland where they first differentiate into epimastigotes and then to the infective metacyclic forms which the fly transmits to the mammalian host. Some variants (designated species) are transmitted mechanically by biting flies (*T. evansi*) or venereally (*T. equiperdum* and *T. equinum*) and have no developmental cycle in the vector (Woo, 1977; Stuart, 1983b). They have kDNA alterations which probably prevent development in the fly, as does the absence of kDNA in dyskinetoplastic mutants of *T. brucei* (Stuart and Gelvin, 1980; Stuart and Feagin, 1992). The mitochondrial respiratory system, components of which are encoded in kDNA, undergoes dramatic changes during the life-cycle of *T. brucei*. Bloodstream forms lack cytochromes and a complete Krebs cycle and generate energy by glycolysis; procyclic forms have a full complement of cytochromes and a complete Krebs cycle and primarily utilize cytochrome-mediated oxidative phosphorylation for ATP synthesis (Vickerman, 1985). Edited mRNAs also exhibit dramatic differences in steady-state abundance between these life-cycle stages and the pattern of these abundance levels parallels the respiratory changes.

2.1 THE MITOCHONDRIAL GENOME OF *T. brucei*

2.1.1 Maxicircles

The maxicircle of *T. brucei* has two distinct regions: the coding region and the variable region (see Fig. 1A). The coding region contains the genes for the two rRNAs and for several proteins (see Jasmer *et al.*, 1987; Simpson *et al.*, 1987; Stuart and Feagin, 1992 for reviews). These are distributed on both DNA strands, although most are on the same strand as the rRNA genes. The rRNA genes have nucleotide sequence homology to other mitochondrial rRNAs, and the amino acid sequences predicted from the DNA or edited mRNA sequences of protein-coding genes have homology to those of other mitochondrial genes. Most of the protein-coding genes encode components of the oxidative phosphorylation system, as is the case in most mitochondrial systems. Genes have been identified for apocyto-chrome b (CYb), cytochrome oxidase subunits I, II and III (COI, II and III) (Benne *et al.*, 1983; Payne *et al.*, 1985; Feagin *et al.*, 1985; Hensgens *et al.*, 1984; Feagin and Stuart, 1985; Feagin *et al.*, 1988a), and mitochondrial NADH dehydrogenase subunits 1, 4, 5, 7 and 8 (ND1, 4, 5, 7 and 8) (Jasmer *et al.*, 1987; Hensgens *et al.*, 1984; Jasmer *et al.*, 1985; Souza *et al.*, 1992; Koslowsky *et al.*, 1990; Payne *et al.*, 1985). Interestingly, the ND7 and ND8 genes have a nuclear or chloroplast location in other organisms; their presence in kDNA in *T. brucei* and other kinetoplastids is the first instance of a mitochondrial genome

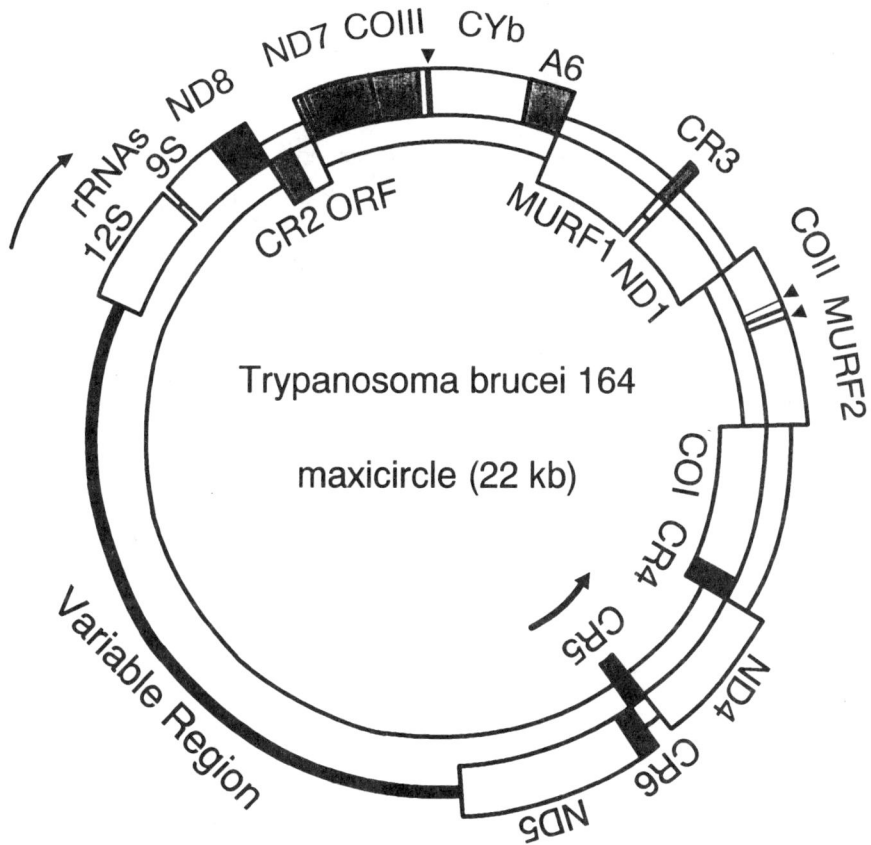

Fig. 1A — The 22 kb mitochondrial maxicircle of *Trypanosoma brucei* that encodes components of the mitochondrial oxidative phosphorylation and protein synthetic systems and contains a non-coding variable region (see text for abbreviations). Sequences encoding edited transcripts are shaded or indicated by arrowheads; the arrows indicate the direction of transcription on each DNA strand.

```
PREPROCESSED CR6/ND5 RNA

DNA...ACCCTTTGTTTTG GTTAA AG A  A ACATCGTTTA  G AAG  AGATTTTAGA ATAAGATATGTTTTTAATATTTTTTTTTATTTTTTTATAATGT...
CR6  .ACCC***G****GtG*TAA AGtAttAtACA*CG**TAttGtAAGttAGA*TTTAGAtATAAGATATGTTTTT(Poly A).
ND5                        5' ATCGTTTA  G AAG  AGATTTTAGA ATAAGATATGTTTTTAATATTTTTTTTTATTTTTTTATAATGT...
PRE...ACCC***G****GtG*TAA AGtAttAtACA*CG**TAttGtAAGttAGA*TTTAGAtATAAGATATGTTTTTAATATTTTTTTTTATTTTTTTATAATGT...
```

Fig. 1B — The maxicircle sequence (DNA) encoding both the 3' end of CR6 mRNA (CR6) and the 5' end of ND5 mRNA (ND5). The nucleotide sequence of a portion of a probable precleaved precursor (PRE) molecule that contains both edited CR6 sequences and ND5 sequences. Uridine residues deleted in the RNA are indicated by asterisks, inserted uridines by lower case 't's.

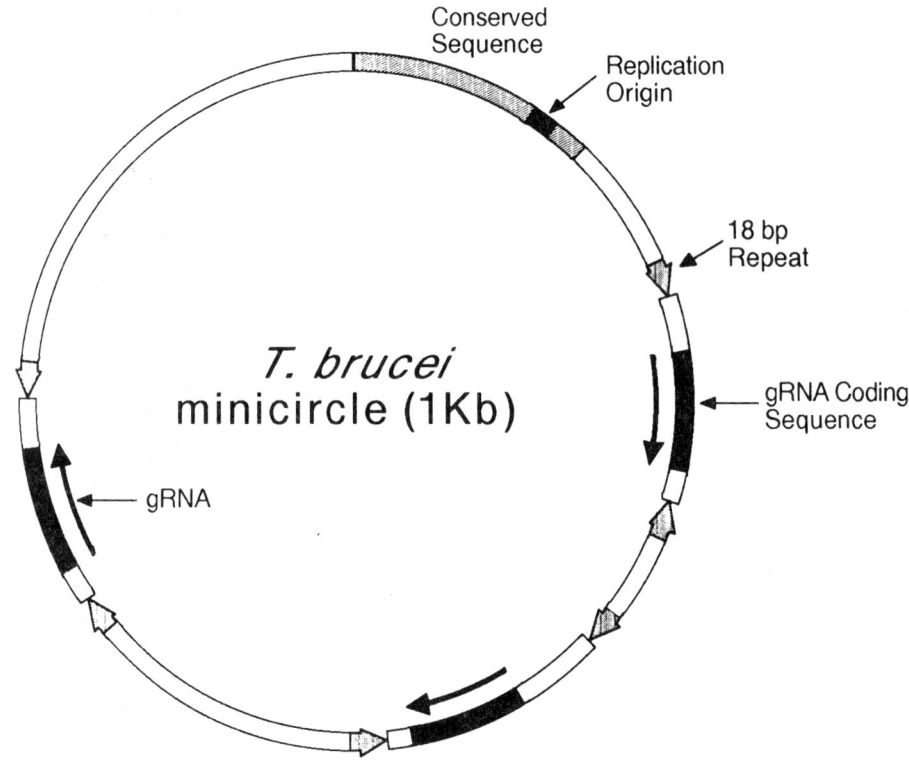

Fig. 1C — A typical *T. brucei* minicircle. The conserved sequence contains the replication origin sequence which is conserved among all kinetoplastid minicircles. The gRNA coding sequences (in black) are flanked by pairs of inverted 18 bp repeats (arrowheads). The gRNAs and their direction of transcription are indicated by the arrows.

location. The maxicircle also appears to encode ATPase subunit 6 (A6) (Bhat *et al.*, 1990) and mitochondrial ribosomal protein S12 (Read *et al.*, 1992b; Maslov *et al.*, 1992; Chapter 3), but see below. The long open reading frames (ORFs) of the regions designated MURF1 and MURF2 suggest that they are protein-coding genes but they have not been identified by homology searches (Payne *et al.*, 1985; Feagin *et al.*, 1985). The genes for ND8 and S12 were initially recognized as guanine-versus-cytosine strand-biased regions and called CR1 and CR6, respectively, like four other such regions named CR2-5 (Feagin *et al.*, 1985; Jasmer *et al.*, 1985; Stuart *et al.*, 1985a). CR stands for cytosine-rich template strand: other workers in the field use the term "G-rich" region to acknowledge the fact that the transcripts are G-rich; see appendix to Chapter 3 for the terminology of RNA editing. The CR2, CR3, CR4, CR5 sequences encode extensively edited RNAs (Stuart *et al.*, unpublished) but their products have not been identified by homology searches. The function of the remaining ORF is unknown; neither is it known whether or not its transcript is edited.

The variable region (VR) of *T. brucei* varies in size among stocks (Stuart and Feagin, 1992). It has a highly repeated sequence organization (Simpson *et al.*, 1987; Stuart *et al.*, 1987; Stuart *et al.*, 1985b; Sloof *et al.*, 1992; Stuart *et al.*, unpublished) and there is no evidence to date that it has a protein-coding function. The region adjacent to the ND5 gene has the highly repeated character unlike the region adjacent to rRNA. There has been little analysis of variable region transcription (Stuart *et al.*, 1987; De Vries *et al.*, 1988; Sloof *et al.*, 1992) and these studies do not reveal abundant discrete transcripts in steady-state RNA consistent with the lack of a protein coding function. No gRNA coding sequences have been found in the VR of *T. brucei*, although they occur at the boundary of this region in *Leishmania tarentolae* (Blum *et al.*, 1990). The significance of the variation of VR sequence and size among species is not known.

2.1.2 Minicircles

Each of the approximately 10,000 minicircles of *T. brucei* is about 1 kb in size (see Fig. 1C). They are quite heterogeneous in sequence but a portion of their sequence and the general sequence organization is conserved among minicircles. Renaturation kinetic analysis indicates that kDNA has a kinetic complexity equivalent to about 300 kb (Stuart and Gelvin, 1980). However, all *T. brucei* minicircles contain a 120 bp conserved sequence and three or four pairs of inverted 18 bp repeats flanking 100 bp non-conserved regions (Jasmer and Stuart, 1986a,b) and thus *T. brucei* kDNA contains about 400 different minicircles. In addition, the renaturation kinetic measurements underestimate minicircle diversity owing to the interspersion of conserved and non-conserved sequences, different abundances of various minicircle sequences, and microheterogeneity. Thus, there arc more than 400 kinds of minicircle per network. This sequence diversity is greater than in other kinetoplastids examined although some laboratory stocks may have lost minicircle sequences (see below). The 120 bp sequence that is conserved among *T. brucei* minicircles has some sequence heterogeneity that includes nucleotide deletions, insertions and substitutions (Jasmer and Stuart, 1986b). This results in the absence of a conserved minicircle ORF, suggesting that it does not have a protein-coding function. The conserved region contains a 12 bp sequence that is absolutely conserved among species and corresponds to a gapped region in replicating minicircles, suggesting that it is the origin of replication (see Simpson, 1987; Ryan *et al.*, 1988). Most *T. brucei* minicircles contain three cassettes composed of the 100 bp sequence blocks and a pair of 18 bp repeats, although some contain four. The sizes but not the sequences of the 100 nt blocks are conserved. gRNAs are encoded in the blocks between the 18 bp inverted repeats (Bhat *et al.*, 1990; Koslowsky *et al.*, 1990). Thus, each cassette appears to contain a gRNA gene and each *T. brucei* minicircle can encode three or four gRNAs. It is possible that gRNAs can be encoded outside the casettes in *T. brucei* minicircles. Other species, such as *L. tarentolae*, encode gRNAs in minicircles but lack the 18 bp repeats. Interestingly, all gRNAs appear to be encoded in the same DNA strand in *T. brucei* minicircles. The single gRNA encoded in the 0.85 kb minicircle in *L. tarentolae* is encoded in the opposite strand (Sturm and Simpson, 1991; Stuart *et al.*, unpublished; see Chapter 3).

2.1.3 Mutants

Mutants of *T. brucei* with altered kDNA occur naturally, and some of these are designated as different species. Mutants have also been produced or occurred spontaneously in the laboratory. The acridine-induced dyskinetoplastic (Dk) mutants of *T. brucei* (Stuart and Gelvin, 1980) and the spontaneous SAK strain of *T. evansi* are devoid of kDNA (Hoare, 1954). The lack of maxicircles and minicircles in Dk mutants is paralleled by the absence of transcripts from both molecules (Feagin *et al.*, 1987). *T. equiperdum* and *T. evansi* (Tobie, 1951) can have partial or complete maxicircle deletions. Interestingly, these mutants retain minicircles but minicircle diversity is dramatically reduced compared with that in the wild type; only one microheterogeneous class of minicircles is retained (Borst *et al.*, 1980). *T. equiperdum* contains both maxicircle and minicircle transcripts and although most, and perhaps all, maxicircle transcripts are not edited, three distinct minicircle transcripts have been identified with gRNA characteristics (Pollard and Hajduk, 1991; Stuart *et al.*, unpublished). However, it is uncertain at this date whether these mutants are capable of editing.

2.2 EDITED RNAs

In *T. brucei* some mRNAs are edited to a limited extent, several others are extensively edited and yet other mRNAs do not appear to be edited at all (see Table 1; for an example of an edited sequence, see Fig. 1B). The COII, CYb and MURF2 mRNAs are edited to limited extents in *T. brucei*. A region near the 3' end of COII mRNA has 4 Us added by editing (Benne *et al.*, 1986; Feagin and Stuart, 1988). This editing eliminates the frameshift encoded in the gene that led to the initial detection of edited RNA. The 5' end of CYb mRNA was examined during an exploration of differential expression of mitochondrial genes during the life-cycle of *T. brucei* (Feagin *et al.*, 1987). This led to the discoveries that this RNA was edited near its 5' end by the addition of 34 Us, that this editing created an in-frame AUG initiation codon, and that the accumulation of edited CYb mRNA was developmentally regulated (Feagin *et al.*, 1987). Non-coded uridines were also found in the poly(A) tail (Stuart and Feagin, 1987) although it is unclear whether these are added by editing. The AUG created by editing in CYb mRNA is within one codon of the AUG created by editing in *L. tarentolae* or *C. fasciculata* (Feagin *et al.*, 1988b), implying that it is functional. In *T. brucei*, however, the created AUG is 20 codons upstream of an encoded in-frame AUG codon for which no corresponding AUG codon is present or produced by editing in *L. tarentolae* or *C. fasciculata*. MURF2 mRNA was analyzed since it also lacks an encoded AUG. It was also found to be edited near its 5' end by the addition of 26 uridines and removal of 4 encoded uridines which creates an AUG, again at the same position as in the other two species (Feagin *et al.*, 1988b).

An important finding was that several RNAs are extensively edited in *T. brucei* as first discovered by studies of COIII mRNA (Feagin *et al.*, 1988a). The COIII gene is not evident in the *T. brucei* maxicircle at the position corresponding to its location in *Crithidia* and *Leishmania*; in retrospect this is an obvious consequence of the extensive editing of COIII mRNA (see Simpson *et al.*, 1987). This region in *T. brucei* has no obvious ORF but has a pronounced guanine-versus-cytosine strand bias. In addition, abundant transcripts from this region are from the C-biased (template) strand, are heterogeneous in size, and larger than expected from the DNA sequence (Feagin *et al.*, 1985). After a COIII mRNA probe from *Leishmania* detected a potential COIII transcript in wild-type but not in the Dk (kDNA deletion) mutant *T. brucei*, RNA and cDNA sequencing was used to determine the consensus sequence of the 969 nt edited COIII mRNA to which 547 uridines are added and 41 encoded uridines are removed by editing (Feagin *et al.*, 1988a; Stuart *et al.*, unpublished). Thus, the sequence of the *T. brucei* COIII mRNA is principally the result of editing.

The sequence upstream of the COIII gene also lacks an obvious ORF, is guanine-versus-cytosine strand biased and is transcribed into abundant guanine biased heterogeneously sized transcripts that are unexpectedly large (Feagin *et al.*, 1985). Determination of the consensus mRNA sequence from RNA and cDNAs (Koslowsky *et al.*, 1990) led to the identification of ND7 mRNA and delineation of the ND7 gene (see below). Interestingly, the ND7 mRNA has two independently edited domains that are separated by a sequence that is not edited and is conserved among *T. brucei*, *L. tarentolae* and *C. fasciculata* (Simpson *et al.*, 1987; see Chapter 3, Fig. 2). The two-domain structure is conserved in *L. tarentolae* and *C. fasciculata* but the edited domains are much smaller in these two species (Shaw *et al.*, 1989; Van der Spek *et al.*, 1988). The sequence downstream of the CYb gene also has the guanine-versus-cytosine strand bias and related characteristics. The consensus sequence determined from RNA and cDNA sequencing reveals an extensively edited mRNA (Bhat *et al.*, 1990). The

Table 1 — Summary of editing in *Trypanosoma brucei*

Gene	Uned.	Us added	Us deleted	Edited	Result	Devel. accum.
12SrRNA	1141	-	-		2-17 Us 3'	
9S rRNA	610	-	-	a	11 Us 3'	
ND8	349	259	46	562	I,T,O	A
CR2	324	345	20	649	I,T,O	A
ORF	226	?	?	?		
ND7	774	553	89	1238	I,T,O	-
ND7-5'	NA	71	13	NA		B
ND7-3'	NA	482	76	NA		A
COIII	463	547	41	969	I,T,O	B
CYb	1117	34	0	1151	I,OE	I
A6	392	447	28	811	I,T,O	B
MURF1	1352	-	-	a		-
CR3	75	148	13	299	I,T,O	A
ND1	1046	-	-	a		-
COII	659	4	0	663	OE	I
MURF2	1089	26	4	1111	I,OE	B
COI	1680	0	-	a		-
CR4	282	325	40	567	?	A
ND4	1358	-	-	a		-
CR5	255	210	13	452	I,T,O	B
CR6	221	132	28	325	I,T,O	A
ND5	1810	-	-	a		A
	total	3583	322			

See text for gene abbreviations. RNA or cDNA sequencing did not detect editing in the 5' end of MURF1, ND1, COI, ND4 and ND5 RNAs or in the unedited portions of CYb or the G- versus-C strand-biased region of COI; S1 nuclease experiments suggest that rRNAs are not edited (a). The remaining RNAs have not been examined for editing. The editing creates initiation (I) and termination (T) codons and the open reading frame (O), or it extends the open reading frame (OE). During the life-cycle the edited RNA preferentially accumulates in the animal (A) bloodstream stages, or in the insect (M) midgut stages, or it occurs at similar levels in both (B) developmental stages.

predicted amino acid sequence has homology to ATPase (A) subunit 6 from other species. This homology is low, which is perhaps not surprising since there is little conservation of the A6 sequence among species. Independent analysis is required to confirm this identification. Nevertheless, much as for COIII mRNA, hundreds of uridines are added and tens are removed from both ND7 mRNAs and A6 mRNAs.

Six other maxicircle regions also exhibit guanine-versus-cytosine strand bias and were named CR1-6; see above (Stuart *et al.*, 1985b). Guanine-versus-cytosinestrand-biased regions occur at corresponding positions in *L. tarentolae* although the sequences are not conserved (Simpson *et al.*, 1987; see Chapter 3). The transcripts from CR1-6 are all extensively edited in *T. brucei* and the edited CR1 transcript appears to be the mRNA for a component of the NADH dehydrogenase complex I (Souza *et al.*, 1992). The mRNA sequence predicts an iron-sulfur-protein that has been named ND8. The edited CR6 transcript encodes a hydrophobic protein that has limited homology to the S12 protein of the small mitochondrial ribosome subunit (RPS12) (Read *et al.*, 1992b; Maslov *et al.*, 1992). The ORF that encodes this protein is conserved between *T. brucei* and *L. tarentolae* and the most conserved regions are those that have homology to RPS12, including domains for streptomycin resistance and streptomycin dependance. However, the very limited homology indicates the need for independent confirmation of its identity, preferably by mitochondrial ribosome location; see Chapter 3. CR2, 3, 4, and 5 RNAs are extensively edited (Stuart *et al.*, unpublished). All appear to encode hydrophobic proteins but as mentioned above these have not been identified by amino acid sequence homology to other proteins, perhaps because their homologous proteins have not been identified yet.

In all cases of extensive editing, the editing creates the initiation and termination codons and a continuous ORF between these codons. The editing in *T. brucei* is more extensive than in *L. tarentolae* and *C. fasciculata* although the same mRNAs are edited. Despite the extensive editing the *T. brucei* COIII, ND7, CR6 and A6 mRNAs have very high nucleotide homology and predicted amino acid sequence homology to the corresponding mRNAs in *L. tarentolae* and *C. fasciculata* as far as these are known. The edited *T. brucei* mRNAs predict proteins that are homologous to those predicted from the less extensively edited mRNAs of *L. tarentolae* and *C. fasciculata* and from mRNAs that are not edited in other organisms. This suggests that the edited transcripts are functional mRNAs which are used for translation.

The 5' ends of COI, COII, ND1, 4, and 5 RNAs are not edited in *T. brucei* (Shaw *et al.*, 1988). The region downstream of the 5' editing domain is not edited in CYb and an internal guanine-versus-cytosine strand-biased region is not edited in COI (Stuart *et al.*, unpublished). There may be small regions of editing that have not been detected owing to their small size and the absence of indicators such as a frameshift in the gene. Nevertheless, editing appears restricted to certain regions of certain transcripts and some transcripts appear not to be edited at all. The editing within the mRNA coding sequences is very precise. The editing at each site is identical among many cDNA clones with minor exceptions (discussed below). Uridines are added within the poly(A) tails of mRNAs and to the 3' ends of gRNAs and rRNAs (Adler *et al.*, 1991; Blum and Simpson, 1990; Campbell *et al.*, 1989; Stuart and Feagin, 1987; Feagin and Stuart, 1989; Benne *et al.*, 1986; Van der Spek *et al.*, 1988). The number and position of Us added within the poly(A) tails is variable, suggesting that these uridines are added by a process that differs, at least to some extent, from that which produces the fully edited mRNAs. The variable number of uridines present at the 3' ends of the gRNAs and the 12S rRNA suggests that these uridines may be added by the activity of the 3' terminal uridylyl transferase (TUTase) that is present in the mitochondrion (see below and Bakalara *et al.*, 1989). The precise number of uridines present at the 3' end of the 9S rRNA (Adler *et al.*, 1991) may be the result of a type of mitochondrial RNA processing that has yet to be characterized.

Some maxicircle genes overlap in the sense that the same maxicircle sequence can specify portions of different mRNAs. The ND7/COIII, COIII/CYb, COII/MURF2, CR4/COI and CR6/ND5 genes are on the same strand and overlap by up to 42 bp of unedited sequence (Read *et al.*, 1992b). For example, the CR6 gene is upstream of the ND5 and the same 37 nt maxicircle sequence that specifies the 3' end of CR6 also specifies the 5' end of ND5 (Fig. 1B). This sequence specifies the 29 nt 5' UTR of ND5 and the downstream 7 nt including the AUG (Read *et al.*, 1992b). This ND5 sequence is not edited (Shaw *et al.*, 1988). The same 37 nt sequence also specifies the 39 nt 3' terminal sequence of CR6 (Read *et al.*, 1992b); the larger size reflects editing in this region of CR6, which creates its termination codon. Interestingly, cDNAs have been characterized which corresponded to potential processing precursor molecules with an edited CR6 sequence upstream of the ND5 sequence. This indicates that editing can precede processing. It further suggests that cleavage of the precursor prior to CR6 editing may produce ND5 mRNA and, alternatively, that editing of the precursor may produce CR6 mRNA. It is tempting to speculate that some mechanism may allow production of both mRNAs from the same precursor molecule, since gRNAs add sequences to mRNAs. The study of mitochondrial RNA processing mechanisms other than RNA editing is likely to be interesting.

2.3 THE GENERAL EDITING PROCESS
The editing of the mRNAs proceeds in the 3' to 5' direction, which was puzzling until the discovery of gRNAs (Blum *et al.*, 1990). The directionality was initially suggested by the characterization of partially edited cDNAs (Abraham *et al.*, 1988; Stuart *et al.*, 1989; Stuart *et al.*, 1990) that are probably derived from RNAs that were in the process of editing when isolated, although this has not been demonstrated by kinetic experiments. Partially edited molecules that are edited in 3' but not in 5' regions are readily detected by Northern blots and analysis of cDNA (see Fig. 2). However, molecules edited in 5' but not in 3' regions are not detected, even in experiments using the sensitive polymerase chain reaction (Abraham *et al.*, 1988). The partially edited cDNAs are edited from the 3' end, for varying distances depending on the cDNA (Stuart *et al.*, 1990). The edited sequences of the numerous cDNAs match each other and match the edited sequence obtained by direct RNA sequencing. In all cases the unedited sequence matches that predicted from the maxicircle gene sequence. The two editing domains of ND7 appear to be edited independently of each

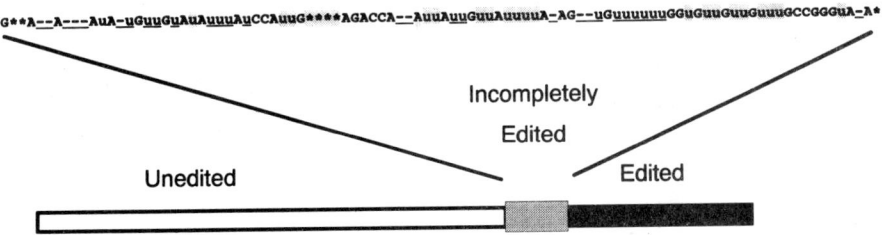

Fig. 2 — Partially edited RNA. The diagram shows the general character of partially edited RNA molecules; these were probably being edited when isolated. These molecules are invariably edited in the 3' but not 5' regions of their editing domains and most have at the junction of the edited and unedited sequences, an incompletely edited sequence; this probably is the site where editing was occurring. In the figure the incompletely edited ND7 sequence shows that sites which require further editing (underlined) are interspersed among those that do not (shaded). The lower case 'u's and '*'s indicate uridines added and removed by editing, respectively, and dashes indicate uridines present in fully edited mRNA.

other, since molecules have been identified that are partially edited in both domains (Stuart *et al.*, 1990). Partially edited molecules are abundant in *T. brucei*, especially for genes whose RNAs are extensively edited. In fact, on the basis of Northern blots, the partially edited mRNAs appear to be substantially more abundant than the fully edited RNAs.

Unedited sequence is immediately adjacent to fully edited sequence in some molecules, but the junction of edited and unedited sequences in most partially edited molecules is incompletely edited in the sense that it does not match fully edited mRNA (see Fig. 2) (Stuart *et al.*, 1990; Koslowsky *et al.*, 1991). This junction region is presumed to be the site at which editing was occurring at the time of RNA isolation. There is one junction region in each partially edited molecule. However, ND7 RNA has two editing domains and can have two junctions per molecule (Stuart *et al.*, 1990). The junctions are at various positions in the different partially edited RNAs, as would be expected depending on how far editing had progressed. The size of the junctions varies, probably also depending on how far toward completion editing had proceeded. The mechanism of the creation of the junction sequences as part of the pathway of RNA editing will be discussed in Sections 2.4 and 2.5.

2.4 gRNAs

The discovery of gRNAs identified the probable source of sequence information for edited RNA. The precise mechanism by which gRNAs specify edited sequence is unknown but the characteristics of gRNAs suggest several likely aspects of their function. The gRNAs contain about 40 nt of encoded sequence

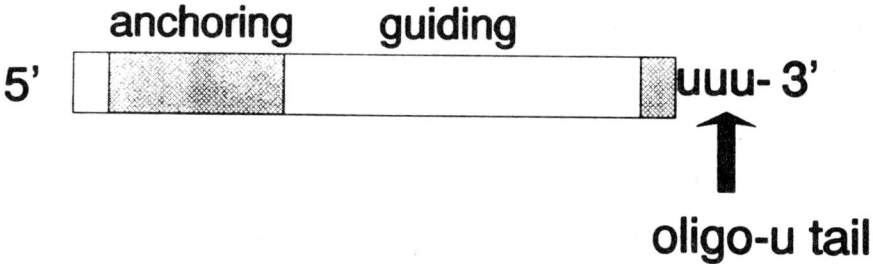

Fig. 3 — Guide RNAs, the apparent specifiers of the edited sequence. Diagram of the general features of the approximately 50-60 nt gRNAs that have regions which can form anchor duplexes with mRNA, guide the edited sequence, and contain non-coded uridine tails.

and an average of about 12 nt of non-coded uridines at their 3' terminus that are presumably added by the TUTase (see Fig. 3) (Bakalara *et al.*, 1989; Blum and Simpson, 1990). The sequences of the gRNAs complement the edited mRNA sequence by a combination of Watson-Crick and guanine-uracil base-pairing. The guanine-uracil base-pairing allows guanine and uracil in gRNA to duplex with cytosine or uracil and adenine or guanine in mRNA, respectively. Hence, the uracils in mRNA can be specified by either guanine or adenine in gRNA. This ambiguity indicates that gRNAs do not act as conventional templates to specify the edited sequence. The match between the 5' end of the gRNA and edited mRNA is almost exclusively Watson-Crick base-pairing, while the remaining match is a mixture of Watson-Crick and guanine-uracil base-pairing. This led to the suggestion that the 5' region of the gRNA forms an initial duplex with pre-edited RNA, called the anchor duplex (Blum *et al.*, 1990). This hypothesis is supported by the finding that only gRNAs that specify the editing of the most-3' region of an editing domain can form an anchor duplex with mRNA before it is edited (Stuart *et al.*, 1989; Koslowsky *et al.*, 1990; Koslowsky *et al.*, 1992b; see Chapter 3). The other gRNAs require editing to create the sequence for the anchor duplex. This suggests that editing by one gRNA produces the sequence that can form an anchor duplex with another gRNA, thus explaining how editing can proceed generally in the 3' to 5' direction as diagrammed in Fig. 4. Several examples have been found which substantiate this model (Fig. 5) (Koslowsky *et al.*, 1992b; Maslov and Simpson, 1992). While numerous gRNAs and gRNA genes have been characterized for *T. brucei*, there is no case where the gRNAs for complete editing of an extensively edited mRNA have been identified. However, as described in Chapter 3, it has been achieved for *L. tarentolae* RPS12 (= CR6) RNA (Maslov and Simpson, 1992).

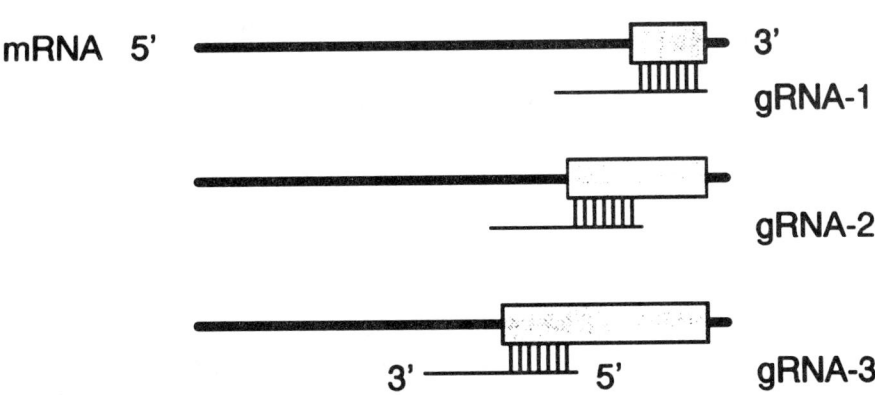

Fig. 4 — Diagram of the proposed sequential utilization of gRNAs that results in editing overall in the 3' to 5' direction. Only gRNA-1 can form an anchor duplex with a target sequence (shaded) in unedited mRNA; the others require editing of the mRNA to form an anchor duplex. The increase in size of the shaded area indicates the progression of the editing process.

Editing appears more complex in *T. brucei* than depicted in the simple model diagrammed in Fig. 4, owing to the occurrence of degeneracy and redundancy in editing. Because of guanine-uracil base-pairing, gRNAs can specify more than one sequence in mRNA (Fig. 5). The consequent microheterogeneity in the mRNA and presumably the resultant protein has been found in cellular RNA (Koslowsky *et al.*, 1990). Conservative amino acid differences resulting from this degeneracy are likely to be tolerated at the physiological level and not be selected against. In addition, different gRNAs can specify the same edited sequence, again because of guanine-uracil base-pairing, and multiple gRNAs that can specify identical edited mRNA sequences have been observed (Stuart *et al.*, unpublished). Furthermore, several gRNAs specify regions that overlap substantially, indicating another type of redundancy. This extensive overlap may represent a form of proofreading.

2.5 gRNA GENES

In *T. brucei*, most of the gRNAs are encoded in minicircles (Koslowsky *et al.*, 1992b; Pollard *et al.*, 1990; Pollard and Hajduk, 1991; Bhat *et al.*, 1990; Koslowsky *et al.*, 1990; Koslowsky *et al.*, 1991). The gRNAs for CYb, MURF2, ND7, and COII were reported to be encoded in the maxicircle in *L. tarentolae* (Blum *et al.*, 1990) and these coding sequences are conserved in *C. fasciculata* (Van der Spek *et al.*, 1991). In *T. brucei*, all CYb and ND7 gRNA genes found

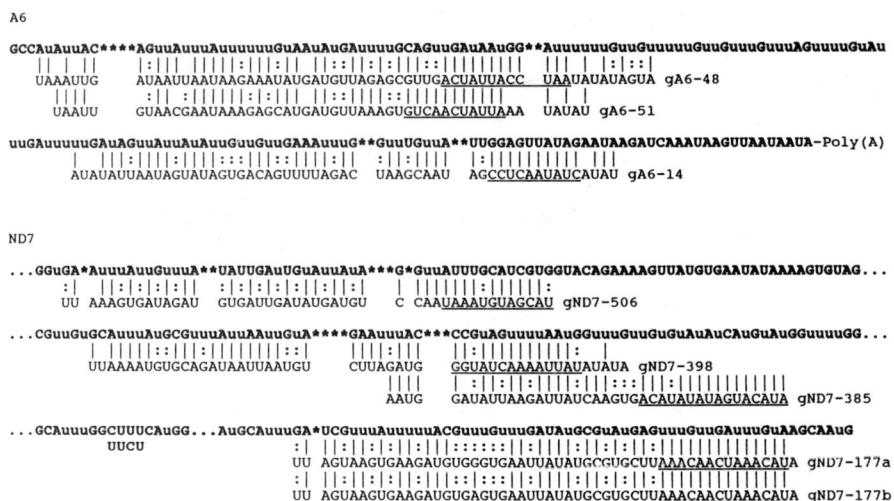

Fig. 5 — Examples of types of gRNA. gRNAs gA6-14 and gND7-506 can form anchor duplexes (indicated by underlining) with unedited RNA at the 3' extremes of A6 mRNA and the 5' editing domain of ND7 mRNA, respectively. gND7-385 can specify the edited sequence with which gND7-398 can form an anchor duplex. gA6-51 specifies editing that substantially overlaps that specified by gA6-48 indicating redundancy which is more extreme in the case of gRNAs gND7-177a and -177b that specify the same edited sequence. The CUUU and UUCU in the bottom line of ND7 mRNA sequence indicate two RNA sequences that are found in mRNA and which can be specified by the same gRNA, indicating the possibility of degeneracy.

to date are located in minicircles (Stuart *et al.*, unpublished) and there is no definitive evidence as yet for any gRNAs encoded in the maxicircle. The maxicircle contains sequences that could encode gRNAs for the editing of the COII frameshift and MURF2 (Van der Spek *et al.*, 1991) but data confirming this function are lacking. The minicircles in *T. brucei* kDNA contain sufficient coding capacity to encode all the gRNAs needed to account for all the editing that occurs in *T. brucei* mitochondrial mRNAs. Each minicircle can encode 3 or 4 gRNAs as diagrammed in Fig. 1B, and thus the kDNA (which has at least 400 different minicircles) can encode more than 1200 gRNAs. Since a total of about 3583 uridines are added by all editing and a gRNA, on average, guides the addition of 15 Us, about 238 gRNAs are required for all the editing observed. Thus, there is ample coding capacity in *T. brucei* minicircles and, in fact, the potential for substantial redundancy.

The significance of the location of the gRNA coding sequences between 18 bp repeats of the minicircles in *T. brucei* is not yet clear. It may reflect their evolutionary history or perhaps have functional significance. Minicircle cassettes may be residues of multiple transposition events that amplified gRNA genes. These genes may have diversified by mutation or recombination. The four-fold repeat structure of the 1.2 kb *T. cruzi* minicircles and the size of the *C. fasciculata* minicircles (2.5 kb), which have no repeats, suggest that they may encode multiple gRNAs. Consistent with its more extensive editing, *T. brucei* has more minicircle sequence diversity than *L. tarentolae* and *C. fasciculata*, although some minicircles may be absent from laboratory stocks of *L. tarentolae*. Genes encoding gRNAs for different mRNAs, as well as for the same mRNA, have been found in single minicircles in *T. brucei* (Koslowsky *et al.*, 1992b; Koslowsky *et al.*, 1991). However, there is no obvious order to the distribution of these genes. The gRNA gene positions do not all correspond to the 3' to 5' order in which gRNAs act and gRNAs for edited mRNAs with different patterns of accumulation between life-cycle stages (see below) can be encoded on the same minicircle.

gRNAs have 5' di- or triphosphate ends as determined by labeling with guanylyl transferase (Blum *et al.*, 1990), a characteristic of primary transcripts. However, it has not been shown whether all gRNAs have such 5' ends and the 5' phosphates may be added posttranscriptionally. The 5' terminal sequence of gRNAs in *T. brucei* is often 5' RYAYA 3', suggesting that this sequence motif plays a role in the initiation of transcription (Pollard *et al.*, 1990) but this has not been directly confirmed. Thus, minicircle transcription requires further analysis to determine whether or not minicircles contain multiple promoters and whether or not there is polycistronic transcription and processing in the production of gRNAs.

2.6 THE POLARITY OF EDITING BY A SINGLE gRNA

While not definitively demonstrated, it is almost certain that gRNAs specify the edited mRNA sequence. However, the precise process by which this occurs is unknown. The striking complementarity between gRNAs and edited mRNA, the detection of covalently linked gRNA/mRNA chimeras as discussed below, and the absence of edited mRNA in mutants lacking gRNA genes is consistent with gRNA's role of specifying the edited sequence. The incompletely edited sequences at the junction of edited and unedited sequence, in partially edited RNAs, such as shown in Fig. 2, have been analyzed since these probably correspond to sites of active editing at the time of RNA isolation.

Initial studies on *L. tarentolae* (Sturm and Simpson, 1990a) described numerous partially edited molecules that shift directly from edited to unedited sequence suggesting that editing proceeded precisely 3' to 5' and led to the hypothesis that the sites of mismatch between gRNA and mRNA were selected for editing (model 1, Fig. 6) (Blum *et al.*, 1990). Molecules with incompletely edited junctions were thought to result from editing with a non-cognate gRNA and evidence for such editing was obtained (see Chapter 3). The extensive editing in *T. brucei* provides an abundance of partially edited molecules for analysis and most of these molecules contain incompletely edited junctions (Stuart *et al.*, 1990; Decker and Sollner-Webb, 1990; Koslowsky *et al.*, 1991).

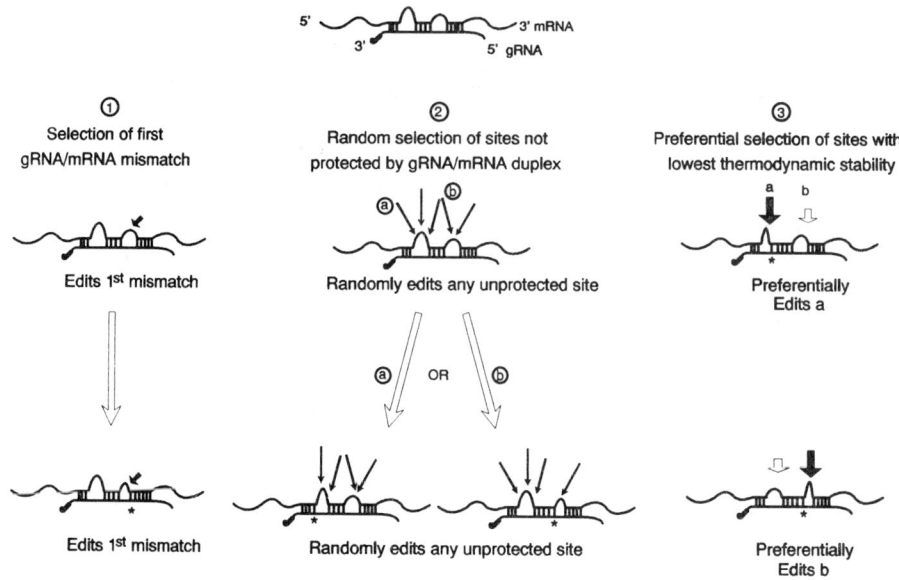

Fig. 6 — Models of editing site selection.
Models for editing site selection which propose that editing sites are selected 1) as the most 3' mismatch of mRNA with gRNA (Sturm and Simpson, 1990a), 2) randomly (Decker and Sollner-Webb, 1990), or 3) preferentially as the least thermodynamically stable region of gRNA/mRNA duplexes (Koslowsky *et al.*, 1991).

The junctions have sites that require editing to match fully edited RNA which are interspersed among those that do not (see Fig. 6). This suggests that sites are edited multiple times to achieve the final RNA sequence. It also suggests that editing does not proceed precisely 3' to 5' within the domain specified by a single gRNA. The detection of incompletely edited chimeric molecules (see Fig. 7) and truncated 5' molecules (see below) strongly supports this conclusion. The detection of incompletely edited junctions led to a second hypothesis that editing sites are selected at random and formation of a correct mRNA/gRNA duplex prevents further editing (Decker and Sollner-Webb, 1990; model 2, Fig. 6).

However, sites with uridines 3' to cytosine are edited. This circumstance is incompatible with both hypotheses, since the U to be deleted would be basepaired with a G, so there is no mismatch that would require further editing. Interactions predicted between the cognate gRNAs and junctions identified substantial regions of duplex, but the alignments differ from those predicted between gRNA and fully edited mRNA (Koslowsky *et al.*, 1991). This led to the hypothesis that regions of lower thermodynamic stability (single-stranded regions) are preferentially selected for editing (Fig. 6, model 3). This accounts for editing other than precisely 3' to 5'. It also suggests that sites are

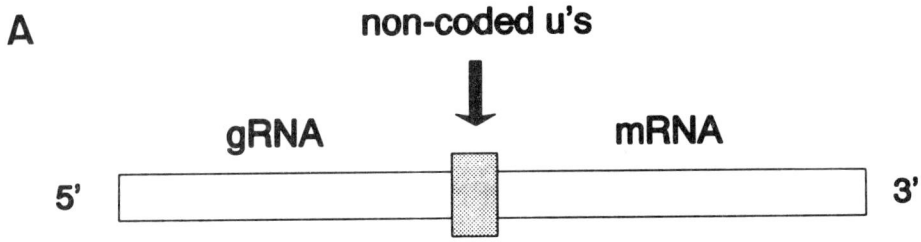

A

non-coded u's

gRNA   mRNA

5'   3'

B

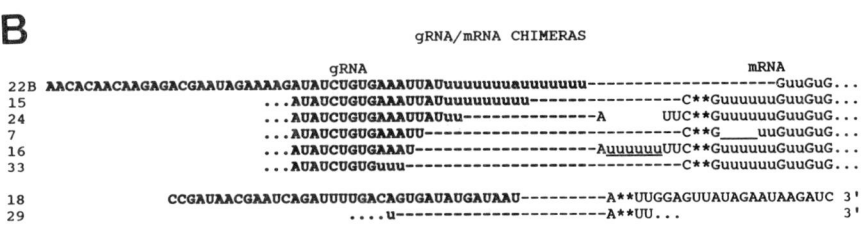

                                    gRNA/mRNA CHIMERAS

                              gRNA   mRNA
22B AACACAACAAGAGACGAAUAGAAAAGAUAUCUGUGAAAUUAUuuuuuuuauuuuuuu-------------------GuuGuG...
15   ...AUAUCUGUGAAAUUAUuuuuuuuuuu----------------C**GuuuuuuGuuGuG...
24   ...AUAUCUGUGAAAUUAUuu--------------A   UUC**GuuuuuuGuuGuG...
7   ...AUAUCUGUGAAAUU-----------------------C**G____uuGuuGuG...
16   ...AUAUCUGUGAAAU-----------------AuuuuuuUUC**GuuuuuuGuuGuG...
33   ...AUAUCUGUGuuu-------------------------C**GuuuuuuGuuGuG...

18   CCGAUAACGAAUCAGAUUUGACAGUGAUAUGAUAAU---------A**UUGGAGUUAUAGAAUAAGAUC 3'
29   ....u--------------------A**UU...   3'

Fig. 7 — Chimeric mRNA/gRNA molecules are likely editing intermediates. (A) Diagram of chimeric molecules showing the relative locations of the gRNA, mRNA and the non-coded uridines. (B) Examples of chimeric molecules from two regions of *T. brucei* A6 mRNA. The gRNA sequences (bold) have 5' extensions that correspond to the PCR primer; the mRNA primer is well downstream and not indicated. Underlined sites are incompletely edited and uridines not encoded in kDNA are in lower case letters.

selected for editing in a preferred but not unique order, owing to dynamic interactions between gRNA and mRNA, and that gRNA and mRNA undergo a series of re-alignments as the mRNA sequence is changed by editing. Thus, reactions proceeding toward complete editing would be thermodynamically favored, which would explain the uridine deletions 3' to C, since these would exchange a G:U pair for a more stable G:C combination.

The hypothesis does not rule out editing by non-cognate gRNAs. The frequency with which this occurs may be influenced by the abundance of such gRNAs relative to cognate gRNAs. The apparent loss of genes for cognate gRNAs in *L. tarentolae* during laboratory cultivation may increase the frequency of editing by non-cognate gRNAs and their detection in this species. For an in-depth discussion of this and related topics the reader is referred to Section 3.15. Conclusive analysis of editing site selection will probably rely on *in vitro* systems to avoid complications arising from the large variety of cellular gRNAs, especially in *T. brucei*.

2.7 CHIMERIC gRNA/mRNA MOLECULES

The proposed role of gRNAs, initially suggested by their complementarity to edited mRNA, was strengthened by the detection of chimeric gRNA/mRNA molecules (Blum *et al.*, 1991; Koslowsky *et al.*, 1991). The gRNAs in chimeras are covalently linked with the cognate mRNA and are linked to the region whose (edited) sequence matches the gRNA (Fig. 7). The 5' truncated molecules that would result from formation of chimeras have also been detected in cellular RNA. Some of these are partially edited at their 3' end (Read *et al.*, 1992a). Importantly, the gRNAs in the chimeras are joined to the mRNA by their 3' uridine tails (Fig. 7). This suggests that these tails are the source of the uridines that are inserted in the mRNA. It also suggests that they may be a repository of the uridines that are removed from mRNA during editing. Surprisingly, the length of the gRNA sequence in chimeras for the same gRNA and editing region is quite variable (Read *et al.*, 1992a; Blum *et al.*, 1991). This suggests that gRNAs of various sizes are produced during or after their transcription before they are utilized to form chimeras, or that they become truncated during chimera formation. It is not yet known whether gRNAs are recycled or whether they are consumed during the editing process. Chimeric gRNA/mRNAs have been formed *in vitro* between synthetic pre-edited mRNA and gRNA that is added or present in mitochondrial extracts (Koslowsky *et al.*, 1992a; Harris and Hajduk, 1992). All requirements for formation of chimeras *in vitro* are not yet known. A 3' hydroxyl on the gRNA is needed, since its elimination by pCp addition or iodination blocks chimera formation, but a 3' uridine is not required. *In vitro* the chimeras resemble those formed *in vivo* since gRNAs are truncated and can be attached at various editing sites.

2.8 THE EDITING MACHINERY

Several activities are required for editing, including RNA cleavage (since editing occurs within the RNA molecules), addition and removal of uridines, and subsequent religation of the RNA. It is likely that these reactions require the participation of molecules other than mRNA and gRNA. Furthermore, the other molecules must have binding sites for the recognition of the gRNA and mRNA substrates. By analogy with other RNA processing mechanisms, especially RNA splicing, the editing complex may be composed of protein and/or RNA in addition to gRNA and mRNA. *In vitro*, experiments have been performed to investigate the potential for specific complex formation with gRNA and pre-edited mRNA (Göringer *et al.*, 1992; Stuart *et al.*, unpublished). As we first reported at the May 1991 Cold Spring Harbor meeting on RNA processing, incubation of either of these radiolabeled RNAs with extracts of mitochondria resulted in the formation of four complexes that are retarded on native gels. Formation of the complexes required Mg^{++} and ATP and modest KCl concentration. The presence of heparin or EDTA or treatment with proteinase or heat inhibited formation of the complexes. Incubation in the presence of molar excess of unlabeled homologous gRNA, mRNA or oligo U but not heterologous tRNA, poly(A) or cloning vector RNA blocked complex formation indicating that complex formation is specific for the gRNA and pre-edited mRNA. Nevertheless, while the optimum conditions for complex formation are similar to those for chimera formation *in vitro*, there is no direct evidence that these complexes form the editing machinery. Crosslinking of the radiolabeled RNA with protein by irradiation with ultraviolet light resulted in the transfer of label to four major proteins with molecular weights of 38, 42, 53 and 96. These represent potential protein components of the editing complex that catalyze the editing reactions.

2.9 MECHANISM OF EDITING

The process by which the final edited RNA sequence is determined is unknown but two general mechanisms are conceivable, as outlined in Fig. 8. One mechanism entails a sequence of enzymatic steps. It proposes that (1) the mRNA editing site is cleaved by endoribonuclease, (2) a gRNA/mRNA chimera is formed by ligation of the 3' terminal uridine of gRNA to the 5' end of the 3' mRNA cleavage product, (3) the chimera is cleaved by endoribonuclease, and (4) the resulting 3' mRNA cleavage product is ligated with the 5' mRNA cleavage product of the first endonucleolytic cleavage. The site of the second endonucleolytic cleavage would determine whether uridines are added or removed. The second mechanism employs the concerted reaction of transesterification that functions in RNA splicing (Cech, 1991). By this process (1) transesterification between the 3' hydroxyl of the gRNA simultaneously creates the gRNA/mRNA chimera and a 5' mRNA fragment and (2) a second transesterification between the 3' OH of the 5' mRNA fragment and the chimera 3' mRNA reconstructs the mRNA. The site of the second transesterification would determine the number of Us added and deleted. The transesterification mechanism has the attractive features that the RNA splicing provides a precedent for its use and that it is energetically conservative, each transesterification accomplishing two enzymatic steps in a single step. However, these models have not yet been experimentally tested.

TUTase, ligase and endoribonuclease activities that are candidates for editing have been found in mitochondria (Bakalara *et al.*, 1989; Simpson *et al.*, 1992; Harris *et al.*, 1992) but they have not been shown to play a role in editing. A likely role for TUTase is to charge the 3' end of the gRNA with uridines; addition of uridines to rRNA (Adler *et al.*, 1991) and the poly(A) tail (Feagin and Stuart, 1989; Campbell *et al.*, 1989; Stuart and Feagin, 1987; Van der Spek, 1988) may be incidental. The presence of endonuclease activity in the mitochondrion is not surprising since such activity may reflect roles in other mitochondrial RNA processing, such as polycistronic mRNA cleavage, RNA stability and possibly gRNA and tRNA processing. Thus, some (and possibly all) ribonucleases may have no role associated with editing. A role other than editing for the ligase is problematical. It might have a role in tRNA 3' end formation but a role in editing is not unlikely. The molecules responsible for the activities have not been purified. Thus, it has not been determined whether the activity is catalyzed by RNA, protein or a heteromeric molecule and whether the activities are obligately associated with a macromolecular complex or able to function as

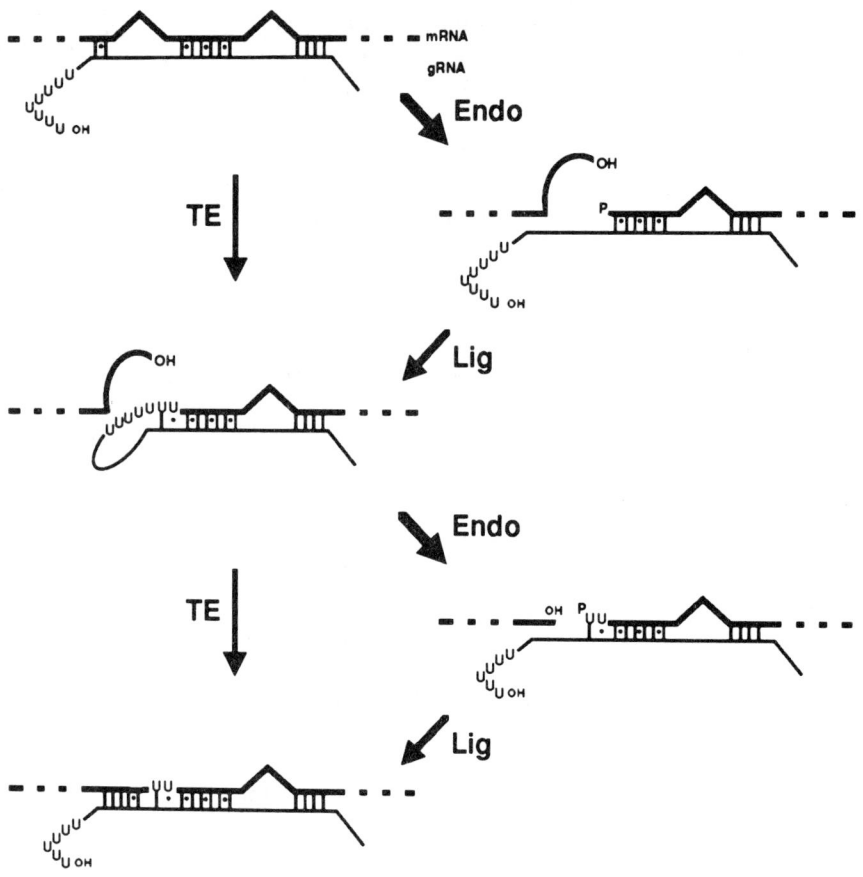

Fig. 8 — The two proposed general mechanisms of editing. The diagrams show that editing can be accomplished by two sequential transesterifications (TE) or two rounds of endoribonuclease cleavage (Endo) and RNA ligations (Lig). The two processes are not necessarily mutually exclusive. A:U and G:C basepairs are indicated by a line, a G:U basepair by a dot. The figure shows the insertion of two Us in an editing site.

separate molecules. Furthermore, the transesterification and sequential enzyme models are not mutually exclusive; both may operate depending on the characteristics of each site being edited.

Irrespective of the mechanism of editing, the means by which a gRNA might specify the edited sequence is unknown. The gRNA cannot be a conventional template as in RNA replication since guanine in gRNA would specify cytosine or uridine in mRNA and uridine in gRNA would specify adenine or guanine. The solution to this problem may reside in that fact that editing appears only to affect uridines in mRNA. However, the mechanism by which this restriction is enforced, remains a mystery.

2.10 DEVELOPMENTAL REGULATION OF EDITING

In *T. brucei* the accumulation of edited mRNA is developmentally regulated. Analysis of this accumulation (summarized in Table 1) is complicated by processes that regulate polyadenylation and RNA abundance. Edited CYb mRNA is abundant in procyclic forms, but slender bloodstream forms have little, if any, edited CYb mRNA (Feagin *et al.*, 1987). Similarly, edited COII mRNA is preferentially present in procyclic forms (Feagin and Stuart, 1988). Stumpy bloodstream forms, non-dividing forms of controversial significance that differ from slender bloodstream forms, have some additional mitochondrial enzymes and some edited CYb and COII mRNA (Feagin and Stuart, 1988). Conversely, ND7 RNA edited in the 3' domain (Koslowsky *et al.*, 1990) and edited ND8 mRNA (Souza *et al.*, 1992) and CR6 mRNA (Read *et al.*, 1992b) are preferentially present in bloodstream forms compared with procyclic forms. Edited A6 and COIII mRNAs are similar in abundance in both bloodstream and procyclic forms (Bhat *et al.*, 1990; Feagin *et al.*, 1988a). Thus, the bloodstream forms accumulate fully edited ND7, ND8, CR6 and A6 mRNA but not edited CYb and COII mRNA, and procyclic forms accumulate edited CYb, COII and A6 mRNA but not fully edited ND7, ND8 and CR6 mRNA.

The level at which this accumulation of edited mRNA is regulated during the life-cycle in *T. brucei* is not known. Both life-cycle stages contain gRNAs that specify the editing, although the monumental task of examining every gRNA for the editing of a particular mRNA has not been performed. Critical gRNAs that specify editing of the most 3' sequences in editing domains are present in stages of the life-cycle where the fully edited RNA does not accumulate (Souza *et al.*, 1992; Koslowsky *et al.*, 1992b). Thus, other gRNAs that specify more 5' sequences could be critical in regulating editing. There are moderate differences in gRNA and pre-edited mRNA abundance and their molar ratios between life-cycle stages. Thus, the accumulation of edited mRNA does not appear to be controlled by the presence or absence of gRNA. It could be controlled by the relative abundance of gRNAs and mRNAs or other factors that control editing, or alternatively by control of edited mRNA stability. The overall abundances of specific maxicircle gene transcripts and of mRNA with longer poly(A) tails are also regulated during the life-cycle (Bhat *et al.*, 1991; Van der Spek *et al.*, 1991). Their abundance parallels the accumulation of edited mRNA (Stuart, 1987; Stuart, 1991) and perhaps similar regulatory processes are employed. The mechanisms that regulate cleavage of pre-edited polycistronic maxicircle transcripts and that add poly(A) tails do not appear to utilize consensus recognition sequences. Perhaps complementary RNAs are utilized in these processes. The mitochondrial structure with the kDNA organized into an network of topologically interlocked circular molecules suggests that some functions such as RNA processing, RNA editing and polyadenylation may be restricted to this organellar region.

2.11 THE ROLES OF EDITING

What might be the function of RNA editing? Editing clearly has the potential to regulate gene expression at the RNA level. It is capable of enabling the translation of RNAs providing them with initiation codons and RNA editing thus may be a translational control mechanism. As mentioned in Section 2.2, unedited CYb RNA contains an initiation codon which occurs in CYb mRNA 20 codons downstream of the AUG created by editing. Translation of unedited CYb RNA might occur but would result in a truncated protein. Similarly, unedited COII and 5' edited ND7 RNAs appear capable of translation but this again would result in the production of truncated proteins. Interestingly, accumulation of edited COII and ND7 RNAs is developmentally regulated. Edited CYb and COII RNAs primarily occur in procyclic forms which have a fully functional mitochondrial cytochrome system. ND7 RNA that is edited in the 5' editing domain is present in both life-cycle stages, but ND7 mRNA that is fully edited in the 3' editing domain is principally present in the bloodstream forms that lack the cytochrome system. Thus, editing has the potential to regulate the composition and abundance of components of the mitochondrial respiratory system. This might be of substantial advantage to an organism that predictably cycles between a life-cycle stage where energy is exclusively generated by glycolysis (bloodstream forms) and a stage where energy is primarily generated by cytochrome-mediated oxidative phosphorylation.

The origin of editing is a mystery. Perhaps it originated in the era of the RNA genome or it may be more recently acquired. An origin in the RNA genome era becomes more attractive as the repertoire of RNA-catalyzed reactions increases. Nevertheless, it seems likely that editing evolved from a more simple form, such as RNA-catalyzed single-uridine addition or removal, into a more complex form. The more extensive editing in *T. brucei* compared with that in *L. tarentolae* and *C. fasciculata* is paralleled by a greater diversity of minicircles, greater gRNA coding capacity per minicircle, and a life-cycle with more dramatic changes in the mitochondrial respiratory system. Interestingly, laboratory stocks of *L. tarentolae* (Souza et al., 1992) and *C. fasciculata* (Arts et al., unpublished) appear not to edit ND8 RNA and other CR sequences, possibly because of the absence of minicircles encoding the cognate gRNAs. These strains grow in culture as the insect-stage promastigotes that employ cytochrome-mediated respiration for ATP synthesis. This suggests that ND8 and perhaps other complex I proteins are not essential for survival in this stage. *C. fasciculata*, some strains of which have very few classes of minicircles, also grow well as promastigotes, and thus may also not require complete complex I. Complex I may be essential to other life-cycle stages of *L. tarentolae* as may have been true for ancestral *C. fasciculata* which has a single life-cycle stage and may have lost life-cycle stages. This suggests that in *L. tarentolae* mitochondrial respiration may be subject to regulation. The more extensive editing in *T. brucei* evolved after its divergence from *L. tarentolae* and *C. fasciculata*. This may have been mediated by amplification of the gRNA genes, perhaps by acquisition of a transposable element now represented by the pairs of

18 bp repeats. Thus, RNA editing probably provides a selective advantage to the parasite.

ACKNOWLEDGEMENTS

I wish to thank B. Smiley and R. Corell for assistance in preparing the figures and to thank the several postdoctoral researchers and technicians who have worked in my laboratory on the mitochondrial genome of *Trypanosoma brucei* and on RNA editing for the stimulating and helpful discussions and exciting experimentation. These studies received support from NIH (AI14102 and GM42118) and I am grateful to The Burroughs Wellcome Trust for the Scholar in Molecular Parasitology Award.

REFERENCES

Abraham, J.M., Feagin, J.E. & Stuart, K. (1988). Characterization of cytochrome c oxidase III transcripts that are edited only in the 3' region. *Cell* **55** 267-272.

Adler, B.K., Harris, M.E., Bertrand, K.I. & Hajduk, S.L. (1991). Modification of *Trypanosoma brucei* mitochondrial rRNA by posttranscriptional 3' polyuridine tail formation. *Mol. Cell. Biol.* **11** 5878-5884.

Bakalara, N., Simpson, A.M. & Simpson, L. (1989). The *Leishmania* kinetoplast-mitochondrion contains terminal uridylyltransferase and RNA ligase activities. *J. Biol. Chem.* **264** 18679-18686.

Benne, R., de Vries, B.F., van den Burg, J. & Klaver, B. (1983). The nucleotide sequence of a segment of *Trypanosoma brucei* mitochondrial maxi-circle DNA that contains the gene for apocytochrome b and some unusual unassigned reading frames. *Nucl. Acids Res.* **11** 6925-6941.

Benne, R. (1985). Mitochondrial genes in trypanosomes. *Trends Genet.* **1** 117-121.

Benne, R., van den Burg, J., Brakenhoff, J.P., Sloof, P., Van Boom, J.H. & Tromp, M.C. (1986). Major transcript of the frameshifted *coxII* gene from trypanosome mitochondria contains four nucleotides that are not encoded in the DNA. *Cell* **46** 819-826.

Bhat, G.J., Koslowsky, D.J., Feagin, J.E., Smiley, B.L. & Stuart, K. (1990). An extensively edited mitochondrial transcript in kinetoplastids encodes a protein homologous to ATPase subunit 6. *Cell* **61** 885-894.

Bhat, G.J., Myler, P.J. & Stuart, K. (1991). The two ATPase 6 mRNAs of *Leishmania tarentolae* differ at their 3' ends. *Mol. Biochem. Parasitol.* **48** 139-150.

Blum, B., Bakalara, N. & Simpson, L. (1990). A model for RNA editing in kinetoplastid mitochondria: "guide" RNA molecules transcribed from maxicircle DNA provide the edited information. *Cell* **60** 189-198.

Blum, B., Sturm, N.R., Simpson, A.M. & Simpson, L. (1991). Chimeric gRNA-mRNA molecules with oligo(U) tails covalently linked at sites of RNA editing suggest that U addition occurs by transesterification. *Cell* **65** 543-550.

Blum, B. & Simpson, L. (1990). Guide RNAs in kinetoplastid mitochondria have a nonencoded 3' oligo(U) tail involved in recognition of the pre-edited region. *Cell* **62** 391-397.

Borst, P., Hoeijmakers, J.H.J., Frasch, A.C.C., Snijders, A., Janssen, J.W.G. & Fase-Fowler, F. (1980). The kinetoplast DNA of *Trypanosoma brucei*: structure, evolution, transcription, mutants. In The Organization and Expression of the Mitochondrial Genome. C. Saccone and A.M. Kroon, eds. (Amsterdam: Elsevier/North-Holland Biomedical Press), pp. 7-19.

Campbell, D.A., Spithill, T.W., Samaras, N., Simpson, A.M. & Simpson, L. (1989). Sequence of a cDNA for the ND1 gene from *Leishmania major*: Potential uridine addition in the polyadenosine tail. Mol. Biochem. *Parasitol.* **36** 197-200.

Cech, T.R. (1991). RNA editing: world's smallest introns? *Cell* **64** 667-669.

Decker, C.J. & Sollner-Webb, B. (1990). RNA editing involves indiscriminate U changes throughout precisely defined editing domains. *Cell* **61** 1001-1011.

De Vries, B.F., Mulder, E., Brakenhoff, J.P.J., Sloof, P. & Benne, R. (1988) The variable region of the *Trypanosoma brucei* kinetoplast maxicircle: sequence and transcript analysis of a repetitive and a non repetitive fragment. *Mol. Biochem. Parasitol.* **27** 71-82.

Englund, P. (1981). Kinetoplast DNA. In Biochemistry and Physiology of Protozoa. M. Levandowsky and S. Hutner, eds. (New York: Academic Press), pp. 333-381.

Feagin, J.E & Stuart, K. (1989). Transcript alteration by mRNA editing in kinetoplastid mitochondria. In Molecular Biology of RNA. T. Cech, ed. (New York: Alan R. Liss), pp. 187-197.

Feagin, J.E., Jasmer, D.P. & Stuart, K. (1985). Apocytochrome *b* and other mitochondrial DNA sequences are differentially expressed during the life-cycle of *Trypanosoma brucei*. *Nucl. Acids Res.* **13** 4577-4596.

Feagin, J.E., Jasmer, D.P. & Stuart, K. (1987). Developmentally regulated addition of nucleotides within apocytochrome b transcripts in *Trypanosoma brucei*. *Cell* **49** 337-345.

Feagin, J.E., Abraham, J.M. & Stuart, K. (1988a). Extensive editing of the cytochrome c oxidase III transcript in *Trypanosoma brucei*. *Cell* **53** 413-422.

Feagin, J.E., Shaw, J.M., Simpson, L. & Stuart, K. (1988b). Creation of AUG initiation codons by addition of uridines within cytochrome *b* transcripts of kinetoplastids. *Proc. Natl. Acad. Sci. USA* **85** 539-543.

Feagin, J.E. & Stuart, K. (1985). Differential expression of mitochondrial genes between life-cycle stages of *Trypanosoma brucei*. *Proc. Natl. Acad. Sci. USA* **82** 3380-3384.

Feagin, J.E. & Stuart, K. (1988). Developmental aspects of uridine addition within mitochondrial transcripts of *Trypanosoma brucei*. *Mol. Cell. Biol.* **8** 1259-1265.

Göringer, H.U., Koslowsky, D.J., Morales, T.H. & Stuart, K. (1992). The characterization of mitochondrial ribonucleoprotein (RNP) complexes involving gRNAs in *Trypanosoma brucei*. *J. Cell. Biochem.* **16A** 136.

Harris, M., Decker, C., Sollner-Webb, B. & Hajduk, S. (1992). Specific cleavage of pre-edited mRNAs in trypanosome mitochondrial extracts. *Mol. Cell. Biol.*

12 2591-2598.

Harris, M.E. & Hajduk, S.L. (1992). Kinetoplastid RNA editing: *in vitro* formation of cytochrome b gRNA-mRNA chimeras from synthetic substrate RNAs. *Cell* **68** 1091-1099.

Hensgens, L.A., Brakenhoff, J., de Vries, B.F., Sloof, P., Tromp, M.C., Van Boom, J.H. & Benne, R. (1984). The sequence of the gene for cytochrome c oxidase subunit I, a frameshift containing gene for cytochrome c oxidase subunit II and seven unassigned reading frames in *Trypanosoma brucei* mitochondrial maxicircle DNA. *Nucl. Acids Res.* **12** 7327-7344.

Hoare, C.A. (1954). The loss of the kinetoplast in trypanosomes, with special reference to *Trypanosoma evansi. J. Protozool.* **1** 28-33.

Jasmer, D.P., Feagin, J.E. & Stuart, K. (1985). Diverse patterns of expression of the cytochrome *c* oxidase subunit 1 gene and unassigned reading frames 4 and 5 during the life-cycle of *Trypanosoma brucei. Mol. Cell. Biol.* **5** 3041-3047.

Jasmer, D.P., Feagin, J.E., Payne, M. & Stuart, K. (1987). Variation of G-rich mitochondrial transcripts among stocks of *Trypanosoma brucei. Mol. Biochem. Parasitol.* **22** 259-272.

Jasmer, D.P. & Stuart, K. (1986a). Conservation of kinetoplastid minicircle characteristics without nucleotide sequence conservation. *Mol. Biochem. Parasitol.* **18** 257-269.

Jasmer, D.P. & Stuart, K. (1986b). Sequence organization in African trypanosome minicircles is defined by 18 base pair inverted repeats. *Mol. Biochem. Parasitol.* **18** 321-331.

Koslowsky, D.J., Bhat, G.J., Perrollaz, A.L., Feagin, J.E. & Stuart, K. (1990). The MURF3 gene of *Trypanosoma brucei* contains multiple domains of extensive editing and is homologous to a subunit of NADH dehydrogenase. *Cell* **62** 901-911.

Koslowsky, D.J., Bhat, G.J., Read, L.K. & Stuart, K. (1991). Cycles of progressive realignment of gRNA with mRNA in RNA editing. *Cell* **67** 537-546.

Koslowsky, D.J., Göringer, H.U., Morales, T. & Stuart, K. (1992a). *In vitro* guide RNA/mRNA chimaera formation in *Trypanosoma brucei* RNA editing. *Nature* **356** 807-809.

Koslowsky, D.J., Riley, G.R., Feagin, J.E. & Stuart, K. (1992b). Guide RNAs for transcripts with developmentally regulated RNA editing are present in both life-cycle stages of *Trypanosoma brucei. Mol. Cell. Biol.* **12** 2043-2049.

Maslov, D.A., Sturm, N.R., Niner, B.M., Gruszynski, E.S., Peris, M. & Simpson, L. (1992). An intergenic G-rich region in *Leishmania tarentolae* kinetoplast maxicircle DNA is a pan-edited cryptogene encoding ribosomal protein S12. *Mol. Cell. Biol.* **12** 56-67.

Maslov, D.A. & Simpson, L. (1992) The polarity of editing within a multiple gRNA-mediated domain is due to formation of anchors for upstream gRNAs by downstream editing. *Cell* **70** 459-467.

Payne, M., Rothwell, V., Jasmer, D.P., Feagin, J.E. & Stuart, K. (1985).

Identification of mitochondrial genes in *Trypanosoma brucei* and homology to cytochrome *c* oxidase in two different reading frames. *Mol. Biochem. Parasitol.* **15** 159-170.

Pollard, V.W., Rohrer, S.P., Michelotti, E.F., Hancock, K. & Hajduk, S.L. (1990). Organization of minicircle genes for guide RNAs in *Trypanosoma brucei*. *Cell* **63** 783-790.

Pollard, V.W. & Hajduk, S.L. (1991). *Trypanosoma equiperdum* minicircles encode three distinct primary transcripts which exhibit guide RNA characteristics. *Mol. Cell. Biol.* **11** 1668-1675.

Read, L.K., Corell, R.A. & Stuart, K. (1992a). Chimeric and truncated RNAs in *Trypanosoma brucei* suggest transesterifications at non-consecutive sites during RNA editing. *Nucl. Acid. Res.* **20** 2341-2347.

Read, L.K., Myler, P.J. & Stuart, K. (1992b). Extensive editing of both processed and preprocessed maxicircle CR6 transcripts in *Trypanosoma brucei*. *J. Biol. Chem.* **267** 1123-1128.

Ryan, K.A., Shapiro, T.A., Rauch, C.A. & Englund, P.T. (1988). Replication of kinetoplast DNA in trypanosomes. *Annu. Rev. Microbiol.* **42** 339-358.

Shaw, J.M., Feagin, J.E., Stuart, K. & Simpson, L. (1988). Editing of kinetoplastid mitochondrial mRNAs by uridine addition and deletion generates conserved amino acid sequences and AUG initiation codons. *Cell* **53** 401-411.

Shaw, J.M., Campbell, D. & Simpson, L. (1989). Internal frameshifts within the mitochondrial genes for cytochrome oxidase subunit II and maxicircle unidentified reading frame 3 of *Leishmania tarentolae* are corrected by RNA editing: evidence for translation of the edited cytochrome oxidase II mRNA. *Proc. Natl. Acad. Sci. USA* **86** 6220-6224.

Simpson, A.M., Bakalara, N. & Simpson, L. (1992). A ribonuclease activity is activated by heparin or by digestion with proteinase K in mitochondrial extracts of *Leishmania tarentolae*. *J. Biol. Chem.* **267** 6782-6788.

Simpson, L. (1986). Kinetoplast DNA in trypanosomid flagellates. *Int. Rev. Cytol.* **99** 119-179.

Simpson, L. (1987). The mitochondrial genome of kinetoplastid protozoa: genomic organization, transcription, replication and evolution. *Annu. Rev. Microbiol.* **41** 363-382.

Simpson, L., Neckelmann, N., de la Cruz, V.F., Simpson, A.M., Feagin, J.E., Jasmer, D.P. & Stuart, K. (1987). Comparison of the maxicircle (mitochondrial) genomes of *Leishmania tarentolae* and *Trypanosoma brucei* at the level of nucleotide sequence. *J. Biol. Chem.* **262** 6182-6196.

Sloof, P., De Haan, A.J., Eier, W.M., Van Iersel, M., Boel, E., Van Steeg, H. & Benne, R. (1992) The nucleotide sequence of the variable region in *Trypanosoma brucei* completes the sequence analysis of the maxicircle component of mitochondrial kinetoplast DNA. *Mol. Biochem. Parasitol.* **56** 289-300.

Souza, A.E., Myler, P.J. & Stuart, K. (1992). Maxicircle *CR1* transcripts of *Trypanosoma brucei* are edited and developmentally regulated and encode a putative iron-sulfur protein homologous to an NADH dehydrogenase subunit. *Mol. Cell. Biol.* **12** 2100-2107.

Stuart, K. (1983a). Kinetoplast DNA: mitochondrial DNA with a difference. *Mol. Biochem. Parasitol.* **9** 93-104.

Stuart, K. (1983b). Mitochondrial DNA of an African trypanosome. *J. Cell. Biochem.* **23** 13-26.

Stuart, K., Feagin, J.E. & Jasmer, D.P. (1985a). Regulation of mitochondrial gene expression in *Trypanosoma brucei*. In Sequence Specificity in Transcription and Translation. R. Calender and L. Gold, eds. (New York: Alan R. Liss, Inc.), pp. 621-631.

Stuart, K., Jasmer, D.P. & Feagin, J.E. (1985b). Organization and expression of mitochondrial DNA in *Trypansoma brucei*. In Achievements and Perspectives of Mitochondrial Research. Volume II: Biogenesis. E. Quagliariello, E.C. Slater, E.F. Palmieri, C. Saccone & A.M. Kroon, eds. (Amsterdam: Elsevier), pp. 337-346.

Stuart, K., Feagin, J.E. & Jasmer, D.P. (1987). Mitochondrial gene expression during development in *Trypanosoma brucei*. In Molecular Stratgies of Parasite Invasion. N. Agabian, H. Goodman & N. Nogueira, eds. (New York: Alan R. Liss, Inc.), pp. 145-155.

Stuart, K., Feagin, J.E. & Abraham, J.M. (1989). RNA editing: the creation of nucleotide sequences in mRNA-A minireview. *Gene* **82** 155-160.

Stuart, K., Koslowsky, D.J., Bhat, G.J. & Feagin, J.E. (1990). The implications of selective and extensive RNA editing. In Parasites: Molecular Biology, Drug and Vaccine Design. N. Agabian and A. Cerami, eds. (New York: Wiley-Liss, Inc.), pp. 111-122.

Stuart, K. (1991). RNA editing in trypanosomatid mitochondria. *Annu. Rev. Microbiol.* **45** 327-344.

Stuart, K. & Feagin, J.E. (1987). Regulation of gene expression in trypanosomes by mRNA editing. In Molecular Genetics of Protozoa. M. Turner, ed. (Cold Spring Harbor: Cold Spring Harbor Laboratory), pp. 143-147.

Stuart, K. & Feagin, J.E. (1992). Mitochondrial DNA of kinetoplastids. In Int. Rev. Cytol.: Mitochondrial Genomes, Wolstenholme, D.R. and Jeon, K.W., eds. (Academic Press, San Diego, CA) Vol. 141, pp 65-88.

Stuart, K. & Gelvin, S.R. (1980). Kinetoplast DNA of normal and mutant *Trypanosoma brucei. Am. J. Trop. Med. Hyg.* **29** 1075-1081.

Stuart, K.D. (1987). Regulation of mitochondrial gene expression in *Trypanosoma brucei. BioEssays* **6** 178-181.

Sturm, N.R. & Simpson, L. (1990a). Partially edited mRNAs for cytochrome b and subunit III of cytochrome oxidase from *Leishmania tarentolae* mitochondria: RNA editing intermediates. *Cell* **61** 871-878.

Sturm, N.R. & Simpson, L. (1990b). Kinetoplast DNA minicircles encode guide RNAs for editing of cytochrome oxidase subunit III mRNA. *Cell* **61** 879-884.

Sturm, N.R. & Simpson, L. (1991). *Leishmania tarentolae* minicircles of different sequence classes encode single guide RNAs located in the variable region approximately 150 bp from the conserved region. *Nucl. Acids Res.* **19**

6277-6281.

Tobie, E.J. (1951). Loss of the kinetoplast in a strain of *Trypanosoma equiperdum. Trans. Am. Microscopic. Soc* **70** 251-254.

Van der Spek, H., van den Burg, J., Croiset, A., van den Broek, M., Sloof, P. & Benne, R. (1988). Transcripts from the frameshifted MURF3 gene from *Crithidia fasciculata* are edited by U insertion at multiple sites. *EMBO J.* **7** 2509-2514.

Van der Spek, H., Arts, G.-J., Zwaal, R.R., van den Burg, J., Sloof, P. & Benne, R. (1991). Conserved genes encode guide RNAs in mitochondria of *Crithidia fasciculata. EMBO J.* **10** 1217-1224.

Vickerman, K. (1985). Developmental cycles and biology of pathogenic trypanosomes. *Br. Med. Bull.* **41** 105-114.

Woo, P.T.K. (1977). Salvarian trypanosomes producing disease in livestock outside sub-Sahara Africa. In Parasitic Protozoa. J.P. Kreier, ed. (New York: Academic Press), pp. 269-296.

3

RNA EDITING IN *LEISHMANIA* MITOCHONDRIA

Larry Simpson, Dmitri A. Maslov and Beat Blum
Howard Hughes Medical Institute, Biology Department and Molecular Biology
Institute, University of California, Los Angeles, CA 90024, USA

3.1 KINETOPLASTID PROTOZOA

The kinetoplastids are flagellated protozoa belonging to the order Kinetoplastida which contains the suborders, Bodonina and Trypanosomatina (Simpson, 1972). Within the family Trypanosomatidae, all of the species are parasitic and are the causal agents of several important human and animal diseases. Cells in the genera *Leishmania*, *Trypanosoma*, *Phytomonas* and *Endotrypanum* are digenetic and go through both a vertebrate and an invertebrate (or plant) host in their parasitic life cycle, whereas cells in the genera *Crithidia*, *Leptomonas*, *Blastocrithidia*, and *Herpetomonas* are monogenetic and only live in an invertebrate host.

All kinetoplastid protozoa possess a single-complex mitochondrion containing a large mass of DNA known as kinetoplast DNA. The kinetoplast DNA consists of thousands of catenated minicircles and a smaller number of catenated maxicircles condensed into a highly organized structure known as the kinetoplast nucleoid body (Simpson, 1987). The precise topology of the minicircles and maxicircles within the nucleoid body is unknown, but the isolated kDNA network is a monomolecular sheet of catenated DNA molecules which appears in the fluorescent microscope after staining with DAPI as a cup-like structure.

The maxicircles are the homologues of the informational DNA molecules in other mitochondria and contain two rRNA genes and at least 11 structural genes encoding proteins mainly involved in electron transport, but no tRNA genes (Simpson, 1987). A schematic representation of the maxicircle gene map of *Trypanosoma brucei* is given in Fig. 1, Chapter 2. For different trypanosome species the order of the genes in the maxicircle and the polarity of transcription is identical (compare Fig. 5).

3.2 THE DISCOVERY OF RNA EDITING

Initial comparisons of the sequenced regions of the maxicircle genomes of *Leishmania tarentolae*, *T. brucei* and *Crithidia fasciculata* resulted in some puzzling findings (Simpson *et al.*, 1987; de la Cruz *et al.*, 1984; Hensgens *et al.*, 1984; Payne *et al.*, 1985). There were no obvious encoded tRNA genes, several of the structural genes lacked methionine-translation initiation codons, and the COII and ND7 genes contained frameshifts within coding regions located at identical relative locations all three (COII) or two (ND7) species. In addition, three genes - COIII, ND7 and MURF4 - were apparently substituted in *T. brucei* by shorter G-rich sequences at the same locations in the genome. Benne and coworkers (Benne *et al.*, 1986) discovered that the frameshift was corrected in the COII gene at the mRNA level by the insertion of four Us at three sites. This phenomenon was termed RNA editing. This discovery was followed by the description of multiple editing events at the 5' portions of several mRNAs which in several cases actually created a methionine-translation initiation codon (Van der Spek *et al.*, 1988; Shaw *et al.*, 1988; Feagin *et al.*, 1988b; Shaw *et al.*, 1989), and finally by the description of extensively or "pan-edited" mRNAs which are edited over the entire length of the gene (Feagin *et al.*, 1988a; Bhat *et al.*, 1990; Koslowsky *et al.*, 1990). (See Appendix for a description of the terminology used in RNA editing.) A summary of all the editing events found to date in mtRNAs in *L. tarentolae* is provided in Table 1.

3.3 PAN-EDITED CRYPTOGENES

The ND7, COIII and MURF4 genes that were present in *L. tarentolae* and substituted in *T. brucei* by shorter G-rich sequences which had no sequence similarity (Simpson *et al.*, 1987) were shown to represent pan-edited cryptogenes (Feagin *et al.*, 1988a; Bhat *et al.*, 1990; Koslowsky *et al.*, 1990). The transcripts of these genes were modified by insertion of multiple Us at multiple sites throughout the length of the RNA, giving rise to mature edited mRNAs which were in some cases twice the size of the gene. As in the case of the internal frameshift edited RNAs and the 5' edited RNAs, no DNA sequences could be found that would correspond to the edited RNA sequences.

3.4 THE MURF4 5'-PAN-EDITED CRYPTOGENE IN *L. TARENTOLAE*

Initially it appeared that pan-editing was limited to the ND7, COIII and MURF4 cryptogenes in the African trypanosome. However, the 5' portion of the MURF4 mRNA in *L. tarentolae* was also found to be pan-edited (Bhat *et al.*, 1990; Bhat *et al.*, 1991). In *T. brucei*, the entire MURF4 mRNA is pan-edited, yielding a

Table 1 — Summary of editing in *L. tarentolae*

Cryptogene	Us added	Us deleted	Sites	Domains
COII	4	0	3	1
ND7	25	0	10	2
CYb	39	0	15	1
MURF2	28	4	12	1
COIII	29	15	17	1
RPS12	117	32	62	3
MURF4	106	5	50	1

protein which shows high similarity with the translated protein from edited *L. tarentolae* MURF4 mRNA.

From a multiple sequence alignment with known ATPase 6 proteins, Bhat *et al.* (1990, 1991) suggested that the MURF4 protein belongs to this family of proteins, although highly diverged (see Section 2.2). However, a Monte Carlo shuffling analysis of the alignments of the MURF4 sequences with known ATPase 6 sequences does not give *Z* values greater than 5 sd units, which is considered the cut-off level for statistically significant sequence similarity (Simpson, L., unpublished results). On the other hand, an analysis using the PROFILE program (Gribskov *et al.*, 1989) yielded somewhat different results (Peris, M. and Simpson, L., unpublished results). First, an alignment of the carboxy terminal domains of the *L. tarentolae* and *T. brucei* sequences was used to calculate a PROFILE. The carboxy terminal portion was used since this is known to be the most conserved portion of ATPase 6 proteins. The PROFILE SEARCH program was then used to scan the Swiss Protein database and the best hits were 12 ATPase 6 proteins, with corresponding *Z* values between 5 and 7.7 sd units. However, a similar search with a PROFILE created from aligned carboxy terminal portions of the human and *Xenopus* ATPase 6 sequences yielded 33 ATPase 6 proteins with *Z* values ranging from 8.0 to 21 sd units. These results are intriguing but do not conclusively show that the MURF4 protein is an ATPase 6 protein. Hydropathy analysis also does not substantiate the claim that MURF4 is ATPase 6 (Peris, M. and Simpson, L., unpublished results). We conclude that the similarity of the kinetoplastid MURF4 amino acid sequences with known ATPase 6 sequences is of limited statistical significance. It is entirely possible that MURF4 represents a highly diverged ATPase 6, but this cannot be confirmed by sequence comparisons alone and, as also concluded in Chapter 2, must rather be substantiated by obtaining the amino acid sequence of the ATPase 6 protein from the kinetoplastids.

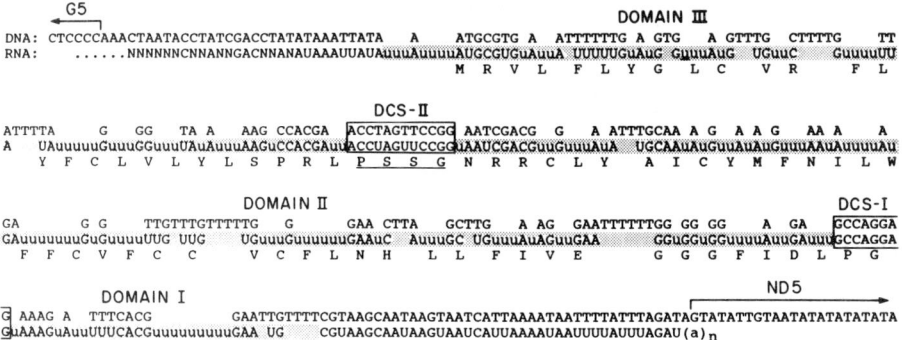

Fig. 1 —— The pan-edited RPS12 (G6) cryptogene from *L. tarentolae*. DNA, RPS12 genomic sequence (nt 14,636 to 14,913 in GenBank entry LEIKPMAX). Adjacent portions and polarity of G-rich region 5 (G5) and NADH dehydrogenase subunit 5 (ND5) gene are shown by arrows. RNA, sequence of mature edited RPS12 mRNA. Editing domains I-III are indicated by shading. Uridines added by editing are shown as 'u's. The amino acid sequence of the translated S12 protein is given beneath the mRNA sequence. Domain-connection sequences (DCS) I and II are boxed. Reprinted from Maslov and Simpson (1992) with permission.

3.5 THE G6 PAN-EDITED CRYPTOGENE IN *L. TARENTOLAE*

It was noted in the initial comparison of the informational regions of the maxicircle genomes of *L. tarentolae* and *T. brucei* that there were six G-rich regions (called CR in Chapter 2; see appendix), which were conserved in location and polarity but not in sequence (Simpson *et al.*, 1987). By analogy with the three G-rich pan-edited cryptogenes (ND7, COIII, MURF4) unique to *T. brucei*, it was suggested that these intergenic sites represented additional pan-edited cryptogenes in both species (Simpson and Shaw, 1989). This hypothesis has been confirmed for the G6 region in *L. tarentolae* (Maslov *et al.*, 1992) and for all G regions in *T. brucei* (Read *et al.*, 1992b; Souza *et al.*, 1992; Stuart *et al.*, unpublished, see Chapter 2).

The mature edited *L. tarentolae* G6 mRNA contains 117 Us inserted at 49 sites and 32 Us deleted at 13 sites (Maslov *et al.*, 1992) (Fig. 1). Editing occurs in three separate editing domains, which are defined as regions of the mRNA which are edited independently of other regions, as evidenced by the 3' to 5' polarity of cDNAs from partially edited molecules. The pan-edited mRNA encodes an 85 amino acid polypeptide which appears to represent a highly diverged ribosomal protein S12. The two most conserved regions of the protein sequences correspond to the sites which are known in *E. coli* to confer streptomycin resistance and streptomycin dependence. Z values of 6.5 - 8.9 sd units were obtained for alignments of the G6 sequence with three chloroplast S12 proteins and one eubacterial S12 protein. Statistically insignificant Z values, however, were obtained for the alignments of G6 with mitochondrial S12 proteins from *Paramecium* and *Zea mays*. The conclusion that the G6 protein belongs to the RPS12 family is strengthened by similarities in hydropathy patterns.

The presence of three editing domains in the RPS12 cryptogene opens the possibility that independent editing of separate domains may modulate amino acid sequences of proteins in addition to simply creating translatable mRNAs. This suggestion must, however, be verified by direct analysis of the proteins translated from edited mRNAs *in vivo*. The other known example of a cryptogene with more than one editing domain is the ND7 pan-edited cryptogene of *T. brucei* (Koslowsky *et al.*, 1990; see Chapter 2). In *L. tarentolae* and *C. fasciculata*, ND7 RNA has an internal frameshift editing site and a 5' editing site, which also may be considered as two separate editing domains, especially since these sites occur at the equivalent of the 5' termini of the two pan-edited domains in *T. brucei* (Shaw *et al.*, 1989) (Fig. 2).

The evolutionary conservation of the G6 regions in *L. tarentolae*, *T. brucei* and *Crithidia oncopelti* could be visualized by deletion of the genomically encoded U residues and alignment of the purine backbones (Maslov *et al.*, 1992). Read *et al.* (1992b) sequenced cDNAs of mature edited G6 (= CR6) RNAs from *T. brucei*. In this species, 132 Us are inserted and 28 Us are deleted, yielding a predicted 82 amino acid protein with good similarity to the *L. tarentolae* RPS12 protein sequence.

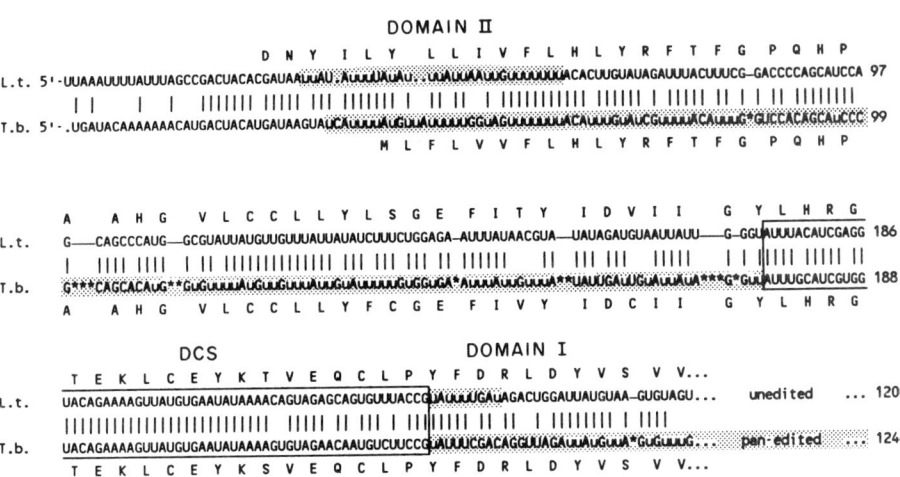

Fig. 2 — Comparison of the editing profiles in mature edited ND7 mRNA sequences from *L. tarentolae* (L.t.) and *T. brucei* (T.b.). GCG GAP program was used to generate the alignment. Matches are shown by vertical bars, periods indicate the gaps introduced by the program. Lower case 'u's indicate uridines added by editing, asterisks indicate uridines which were deleted. Editing domains are shaded and the DCS is boxed. The amino acid sequences of the translated ND7 proteins are given above and below the corresponding mRNA sequences. The T.b. edited RNA sequence is from Koslowsky *et al.* (1990).

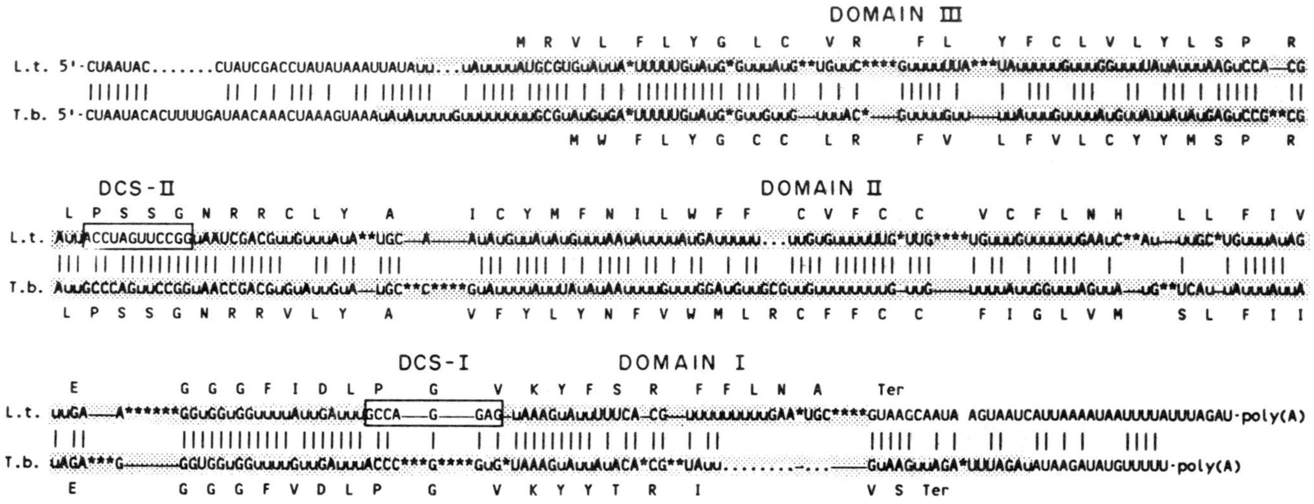

DOMAIN III

DOMAIN II

DCS-I

DOMAIN I

DCS-II

Fig. 3 — Editing domains in the RPS12 cryptogene are not conserved between species. The amino acid sequences of the translated S12 proteins are given above and below the corresponding mRNA sequences from *L. tarentolae* (L.t.) and *T. brucei* (T.b.). See Fig. 1 and 2 legends for other designations. The T.b. sequence was taken from Read *et al.* (1992b).

Alignment of the translated RPS12 amino acid sequences and the mature edited RNA sequences of the *L. tarentolae* and *T. brucei* genes shows that the two domain-connection sequences (DCSs), which represent the functionally signficant regions of the protein, show absolute sequence conservation at the amino acid level (Fig. 3). However, at the RNA level, the DCSs are edited in *T. brucei*. This eliminates the possibility that the existence of the unedited DCSs in *L. tarentolae* is due to a functional constraint on editing in those regions.

Productive pan-editing of the transcript from the G1 (= CR1) region in *T. brucei* has also recently been demonstrated (Souza *et al.*, 1992). The encoded protein shows sequence homology with an iron-sulfur component of bovine NADH dehydrogenase. Attempts to detect pan-edited G1 RNAs from *L. tarentolae* and *C. fasciculata* (Souza *et al.*, 1992; Arts *et al.*, unpublished) by PCR amplification were unsuccessful.

3.6 DISCOVERY OF GUIDE RNAs

The mystery of the missing sequence information for the edited RNAs was solved by the discovery of guide RNAs (gRNAs) (Blum *et al.*, 1990). These are small RNAs which can form perfect hybrids with mature edited mRNAs if G-U base pairs are allowed. The region of base complementarity extends 3' of the pre-edited region for a variable length. Formation of this 3' anchor duplex is thought to represent the initial interaction between a specific gRNA and a

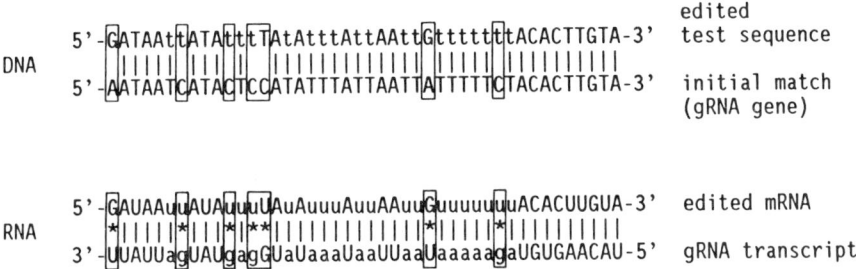

Fig. 4 — Discovery of gRNA genes. The GCG BEST FIT program was used with the edited ND7 test sequence to identify the putative gRNA gene antisense sequence in the LEIKPMAX maxicircle sequence. The default weight matrix which scores identities was used. The resulting characteristic G-A and T-C mismatches in the initial alignment are boxed. These mismatches were resolved by using the putative transcripts of the DNA sequences, allowing for G-U base pairs (boxed) (Blum *et al.*, 1990).

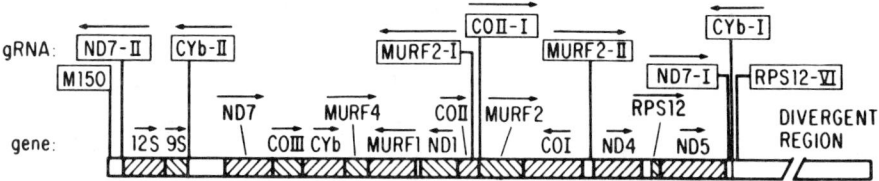

Fig. 5 — Localization of gRNA genes on the genomic map of the *L. tarentolae* maxicircle. The circular map was linearized at the single EcoRI site and the divergent region is not to scale. The structural genes are indicated by cross-hatching and the polarities shown by arrows. The designations for gRNA genes are boxed with the polarities given by arrows. RPS12 corresponds to G-rich region 6 (G6). The position of the other G-rich regions has not been indicated. Reprinted (with modifications) from Blum *et al.* (1990), with permission.

specific pre-edited mRNA to initiate the editing process (Blum *et al.*, 1990). The gRNAs contain heterogeneous 3' oligo-[U] tails (Blum and Simpson, 1990) and can easily be recognized on an acrylamide-urea gel by a characteristic multiple banding pattern migrating ahead of tRNAs (Blum *et al.*, 1990; Sturm and Simpson, 1990b; Blum and Simpson, 1990). gRNAs from maxicircle (and minicircle) DNA were actually originally visualized on Northern blots as contaminants of *L. tarentolae* mitochondrial tRNA preparations, but the significance of this observation was not appreciated at the time (Simpson *et al.*, 1989).

The existence of gRNAs was predicted by a computer analysis of the known 21 kb *L. tarentolae* maxicircle DNA sequence, which was searched for short sequences capable of forming perfect hybrids with regions of edited mRNAs (Blum *et al.*, 1990). No such hybrids were initially found, but we noted that the mismatches were characteristically G-A and T-C. If G-U base pairs were allowed, then the transcripts of these DNA sequences would form perfect hybrids with edited mRNAs (Fig. 4). G-U is a weak but bona fide base pair that occurs in tRNAs and rRNAs. The presence of G-U base pairs excludes the participation of a standard polymerase activity in catalyzing the transfer of information from the gRNA to the mRNA.

The presence of the gRNAs was established by hybridization of Northern blots of kRNA using synthetic oligomer probes for the predicted sequences and by direct 5' RNA sequence analysis (Blum *et al.*, 1990). Three gRNAs (gCYb-II, gMURF2-II, and gND7-II) were isolated by hybrid-selection revealing a non-encoded oligo-[U] tail at the 3' end (Blum and Simpson, 1990). One of the hybrid-selected gRNAs (gND7-II) was shown to contain a di- or tri-phosphate at the 5' end, suggesting that gRNAs represent primary transcripts (Blum and

Simpson, 1990). Initially, seven maxicircle-encoded gRNAs for four of the five known cryptogenes were discovered and shown to exist (Blum *et al.*, 1990) (Figs. 5, 6). The gRNA sequence for the editing of the COII frameshift was located *in cis* within the mRNA at the 3' end, whereas all other gRNAs were *in trans*.

Van der Spek *et al.* (1991) showed that all seven putative gRNA genes were conserved in equivalent locations in the maxicircle genome of *C. fasciculata*. The presence of compensatory base changes which preserved base-pairing of the anchor regions of the mRNAs and the gRNAs in this species provides strong evolutionary evidence for the participation of gRNAs in RNA editing. In *T. brucei*, only the gMURF2-I, gCOII and gMURF2-II gRNAs were present in equivalent maxicircle locations. The gND7(5')-II, gCYb-I and -II, and gND7(FS)-I gRNA genes could not be identified in the *T. brucei* maxicircle sequence.

3.7 DISCOVERY OF THE GENETIC ROLE OF MINICIRCLE DNA

The absence of a gRNA in the maxicircle for mediating the editing of the COIII RNA led to a search of minicircle DNA sequences, of which there were at that time three complete and several partial sequences. It was also known from our previous work that minicircles encoded small transcripts with the characteristic mobility on acrylamide gels of gRNAs (Simpson *et al.*, 1989). Computer analysis indicated a putative minicircle-encoded gRNA (the D12 minicircle) for editing of the first eight sites of the COIII mRNA and this was shown to exist by direct RNA sequencing (Sturm and Simpson, 1990b). Shortly thereafter, the kDNA minicircle molecules of *T. brucei* and *T. equiperdum* were shown to each encode three putative gRNA-like transcripts (Pollard *et al.*, 1990; Pollard and Hajduk, 1991). The presence of a 5' terminal di- or triphosphate for the gRNAs from all three species suggested that these were primary transcripts. The gRNAs in the African trypanosomes are situated between 18mer inverted repeats within the variable region of the minicircle (Pollard *et al.*, 1990; Bhat *et al.*, 1990; Koslowsky *et al.*, 1990) (Fig. 7). In *L. tarentolae*, there is a single gRNA gene per minicircle located within the variable region approximately 150 bp from the end of the conserved region, and the 18mer inverted repeats are not present (Sturm and Simpson, 1990b) (Fig. 7).

The discovery of gRNA genes within minicircles provided the first evidence for a genetic role of minicircle DNA. It is entirely possible, however, that minicircle DNA has additional functions, such as, for example, in determining the structure of the nucleoid body *in situ* and mediating the segregation of the maxicircle molecules to daughter mitochondria.

3.8 COMPLETE SETS OF OVERLAPPING gRNAs FOR COIII, MURF4 AND RPS12

Two plasmid minicircle libraries were constructed for *L. tarentolae*, and clones for new gRNA sequence classes were selected by a subtractive hybridization protocol (Table 2). Mixed oligonucleotide probes for the variable regions of the five known minicircle sequence classes were used to select clones of the known sequence classes. The remaining clones were sequenced and a total of 12 new minicircle sequence classes were identified in the two libraries (Maslov and Simpson, 1992). Potential gRNA genes were identified at equivalent locations in the variable regions in each new minicircle, and the gRNAs identified by primer extension sequencing. The functional role of each of the new gRNAs was determined by sequence comparisons with known edited mRNA sequences (Maslov and Simpson, 1992).

As shown in Table 2, the X4 minicircle encodes the gCOIII-II gRNA for the remaining nine sites of the COIII pre-edited region. Five of the new minicircles (and, in addition, the previously identified D3 minicircle) encode gRNAs for the 5' pan-edited MURF4 mRNA. The remaining six new minicircles (and, in addition, the previously identified Lt154 minicircle) encode gRNAs for the pan-edited RPS12 mRNA. The remaining gRNA for this gene was identified as a maxicircle-encoded transcript (gRPS12-VI) by computer analysis and direct RNA sequencing (see Fig. 5). The MURF4 mRNA in *L. tarentolae* is edited at 50 sites by the addition of 106 Us and the deletion of 5 Us. This entire editing scenario is mediated by six overlapping minicircle-encoded gRNAs (Fig. 8A). The precise overlapping of gRNAs provides an explanation for the observed 3' to 5' polarity of editing within an editing domain, in that the upstream anchor sequences are created *de novo* by downstream editing.

Editing of the RPS12 mRNA in *L. tarentolae* is more complex and involves three separate domains (Fig. 8B,C). Seven minicircle-encoded gRNAs and one maxicircle-encoded gRNA mediate editing of this mRNA. As in the case of COIII and MURF4, the 3' to 5' polarity of editing within a domain is determined by the creation of upstream anchor sequences by editing. However, the initial gRNA for each domain forms an anchor with an unedited mRNA sequence.

Examination of the gRNA/mRNA hybrids formed in MURF4 and RPS12 pan-editing indicates that the anchor duplexes have relatively few G-U base pairs compared with the gRNA/edited mRNA duplexes. An examination of the relative thermodynamic stabilities of the duplex regions formed between the 5' end of the gRNA and the mRNA anchor sequence, and between the 3' end of the downstream gRNA and the edited mRNA sequence, indicates that a role for the relatively weak G-U base pairs is to destabilize the downstream duplex and to allow formation of the upstream anchor duplex with the adjacent gRNA (Maslov and Simpson, 1992). The ratio of the free energy of formation of the anchor duplex versus the gRNA-edited mRNA duplex for the RPS12, MURF4 and COIII genes varied from 1.3 to 2.9. This, of course, does not eliminate the possibility of an RNA helicase which could destabilize the gRNA-edited mRNA duplex at the site of the new anchor. The need for a helicase is more apparent in the case of the 5' terminal editing block in a pan-edited gene, or in the case of a single block of editing which is mediated by a single gRNA. This helicase activity might actually be provided by the ribosome binding to the putative ribosome-binding site which could be created by the 5' terminal editing events (Pause and Sonenberg, 1992).

Complete sets of overlapping gRNAs

Table 2 — Analysis of minicircle plasmid libraries

Library I. XmaI-cut minicircle clones (2,500 clones)
Negative selection with five classes: Lt19, B4, Lt154, D3 and D12

Number of selected clones	Number of analyzed clones	Minicircle sequence class	Number of clones	Encoded gRNA
Old classes:				
		Lt19	59	gLt19
		B4	28	gB4
2,400[a]	92[b]	D3	3	gMURF4-III
(96%)		D12	3	gCOIII-I
		Lt154	0	gRPS12-IV
New classes:				
		X1	28	gRPS12-III
		X2	1	gMURF4-VI
		X3	3	gMURF4-I
		X4	9	gCOIII-II
		X5	10	gRPS12-I
100[c]	100	X6	31	gRPS12-V
(4%)		X7	7	gMURF4-V
		X8	1	gRPS12-II
		X9	2	gMURF4-II
		X10	3	gRPS12-VIII
		X11	3	gRPS12-VII

Library II. (MspI+ethidium bromide)-cut minicircle clones (~25,000
clones). Negative selection with all above classes except X2 and X8.

34[d]	34	X2	24	gMURF4-VI
		X8	7	gRPS12-II
		M1	3	gMURF4-IV

[a]The clones were positive for the five known minicircle classes.
[b]Randomly chosen clones were analysed.
[c]The minicircle DNA clones which were negative for the five known sequence classes.
[d]The minicircle DNA clones which were negative for fourteen other sequence classes.
Reprinted from Maslov and Simpson (1992) with permission.

COII

```
                        gCOII-I
                UUAUAUCUaaUaUaUUGGAUAUAUU...
                || :||||||:||||||:  :    (103nt)
                            3   2  1
5'·····AUG...(490nt)...GAGUUAAAGUAGAUuGUAUACCUGGUCGC...·
```

MURF2

```
                                            gMURF2-II
   [U]n-U—U—UaUaaaUUaaUaaaUUaaaUaUagaaUaUaUgaaaCUGACUCUGAUUA-5'
   | | ||:|||::||:|||||||||::||||||||::|||||||  ::::   gMURF2-I
                                        UUACaUUAGUACAUA-5'
                                        | |||+|||+|||||
5'...UUU...(22nt)...UAUUA**A**A*AuGuuuuGGuUGuuuuAAuuuAGuuuuAuuuuuAuGuuuuGACUGuAGuUCGUGUAUUUGAUU...(1004nt)...UAA
   12 11 10 9    8 7 6    5 4    3    2     1
```

CYb

```
                                           gCYb-II
   [U]n-UUUUUaUaUaaaagaGUaCaaUUUggaaaCCaaUgaaaaaaaUaaUaaaUCUUUUC-5'
   | |||||||:||||:||:|||||:|:||||||||||||
                                                 gCYb-I
                         UUUUaaaUaUaaaUaGaAAUUGAAGUUCAGUAUAUACAC-5'
                         :|||||||:||:|||||||||||||+|[:: |
5'·AUA...(20nt)...UAAUUAUAAAuAuGuuuuuuCGuGuuAGAuuuuuuGuuAuuuuuuuAuuAuuuuAGAAAuuuuAuGuuUGuuGuCuUUAACUUCAGGUUGUUAUUUACGAG...(1028nt)...UAA
   ^ 15 14 13  12 11   10   9   8   7 6   5 4   3 2 1
```

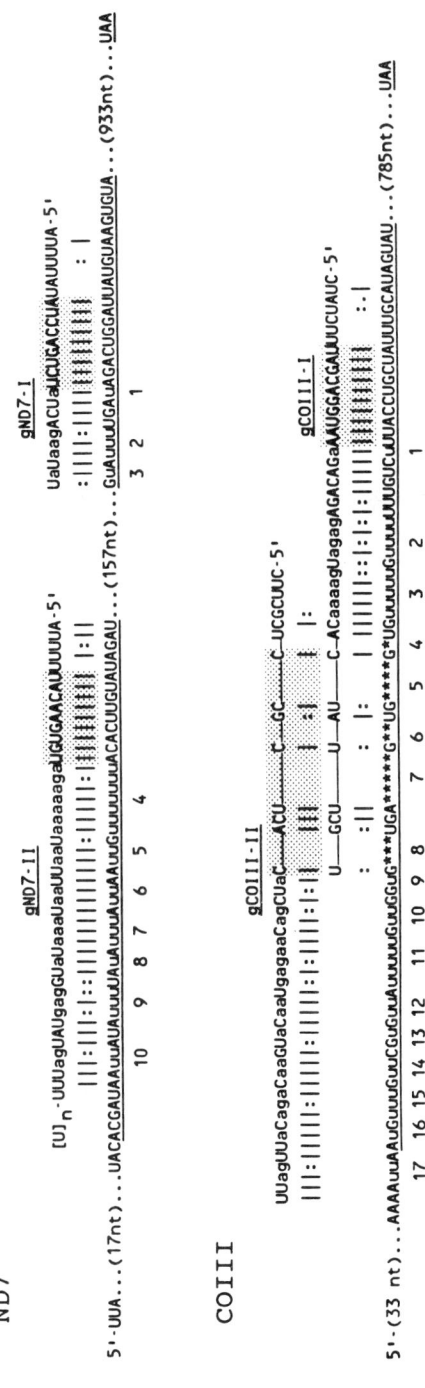

Fig. 6 — Alignments of edited RNA sequences of several mitochondrial mRNAs with corresponding gRNA sequences. The added uridines in the mRNA are indicated by 'u's and the deleted uridines are shown by asterisks. The open reading frames are underlined and edited sites are numbered 3' to 5'. Guide nucleotides in gRNA sequences are indicated as lower case 'a's and 'g's. The position of the oligo-[U] tail is indicated by [U]n for the two gRNAs in which the 3' ends were sequenced directly. The 3' anchor duplex regions between gRNA and mRNA are shown by shading. Canonical A-U and G-C base pairs are indicated by vertical bars and G-U base pairs by colons.

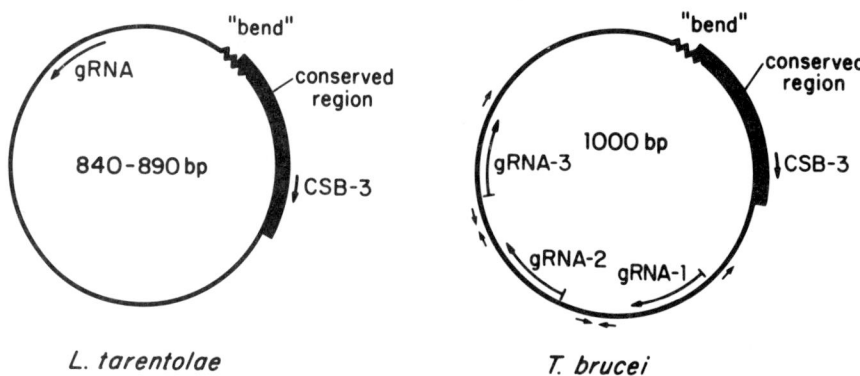

Fig. 7 — Genomic organization of kDNA minicircle molecules from two species. The locations of the conserved region, the DNA "bend", and the gRNA genes are indicated. Note that the polarity of the gRNA genes in *L. tarentolae* (Sturm and Simpson, 1991b, c) and *T. brucei* (Pollard *et al.* 1990; Pollard and Hajduk, 1991; see Fig. 1C of Chapter 2) is opposite. CSB-3 (conserved sequence block 3) is a 12-mer involved in replication initiation. Small arrows indicate the 18mer inverted repeats flanking the *T. brucei* gRNA genes.

A

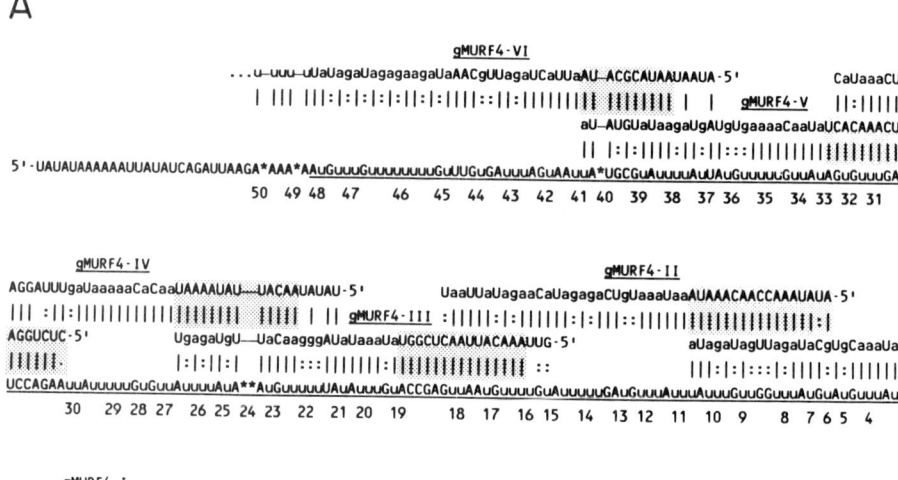

Fig. 8 — Multiple overlapping gRNAs mediate editing of MURF4 mRNA (A) and RPS12 mRNA (B). DCS-I and DCS-II are boxed. See legend to Fig. 6 for other symbols. Reprinted from Maslov and Simpson (1992) with permission. (C) Diagrammatic representation of the editing of the RPS12 mRNA in *L. tarentolae*.

B

gRPS12-VIII

UAUagaUaagaUAUGCACaUaaU~GGAGACaUAC~~CAAAUAC~~~ACAUC~~~~UC-5'

||||:||||:|||:||||||||| ::|:|||||| ||||||| ||| : gRPS12-VII

  aU~UagaUaC~GUaaG~~UagaaGAU~~AUaaaaa

  |: :|:|||| ::||| :|:||:|| |||||||

5'-CUAAUACCUAUCGACCUAUAUAAAUUAUAuuuAuuuuAUGCGUGuAuuA*UUUUUGuAuG*GuuuAuG**UGuuC****GuuuuUUA***UAuuuuu

   62 61   60 59 58 57 56- 54 53 - 51 50 49   48 47

gRPS12-VI

UaaaUUaaaAUaUaagUUCaGGUGUUaaUGGAUUAAGGUAUUAUUA-5'   gRPS12-IV

:|||::|||||||||||||||||| ||||||||:||||: | aUaUaaaUUgUgagaUaCUaaagagaUa

CAAACCAAAAUAUAG-5'   gRPS12-V |||:|||||:|:|:||||||||:|:|

||||||||||||||:   UUaUUAGUUGCaaCagaUaU~~ACGUGaUaCgaUAUACAAAUUUAAGAUAC-5'

          ::|||||:||||||:||| |||| ||||:||||||||| |:|

GuuuGGuuuUAuAuuuAAGuCCACGAuuACCUAGUUCCGGuaaUCGACGuuGuuuAuA**UGCAAuAuGuuAuAuGuuuAAuAuuuuAuGAuuuuuuGu

 46 45 44 43 42   41 DCS-II 40   39 38 - 36   35 - - - - 30 29 28 27 26 25

                                gRPS12-II

CaAAAAAC~AAC~~~~~~ACAAAAUA-5'   ACagaUgUUggCUC~~~~~~UCaCUaUUCaagaUaaUUagaCGGUCCUCAACUA-5'

||||||||| ||| ||||| | gRPS12-III |||:||:|:|:::||. :|||:|:|||:||||:||:|||||||| .

aaaGGC~GAU~~~~~GUaaaUaaagagCUUaG~~UagaCG~ACAAAUGUCACCUA-5'   UUaUUUCaCaaG

|||::| :|:   ::|||:|||:|:||||| ||:||| ||||||:||| || |:||||||:||

GuuuuUUG*UUG****UGuuuGuuuuuuGAAuC**AuuuGC*UGuuuAuaGuuGAA******GGuGGuGGuuuuAuuGAuuuGCCAGGAGuAAAGuAuuU

24 23 22 21 20 19 18 17 16 15 14 13   12 11 10 9 8 7 DCS-I 6 5 4

gRPS12-I

AAGUUCagaaggagaCUU~AC~~~~GCAUUCGCUUCUC-5'

|||| ||:||::|:|||| || ||||||||-| -||

UUCACGuuuuuuuuuGAA*UG****CGUAAGCAAUAAGUAAUCAUUAAAAUAAUUUUAUUUAGAU-poly(A)

   3   2   1

C

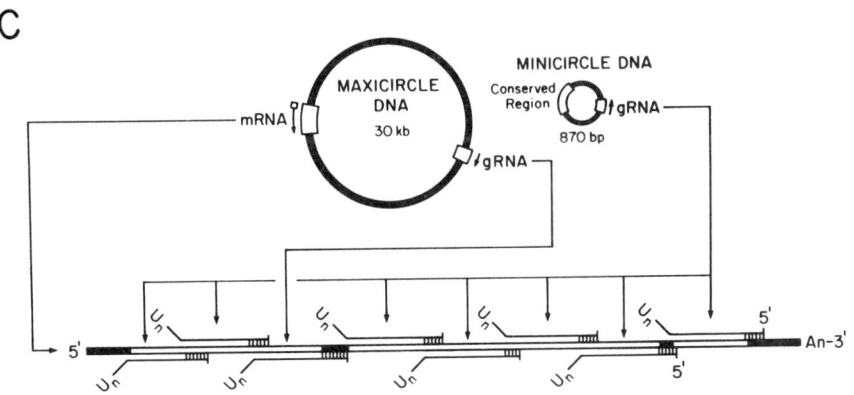

Table 3 —— The genomic complexity of gRNAs in *L. tarentolae* strain UC

Cryptogene	Assigned gRNAs[a]		Putative gRNAs	
	Number of gRNAs encoded by		Putative	Expected number
	maxicircle DNA	minicircle DNA	cryptogene	of gRNAs
COII	1	-	G1	12
ND7	2	-	G2	12
CYb	2	-	G3	4
MURF2	2	-	G4	8
RPS12	1	7	G5	7
COIII	-	2		
MURF4	-	6		
Total	8	15		43

[a]Three unassigned gRNAs were identified: gLt19 and gB4 (minicircle-encoded) and gM150 (maxicircle-encoded). Reprinted from Maslov and Simpson (1992) with permission.

3.9 COPY NUMBER OF MINICIRCLE SEQUENCE CLASSES

The relative abundance of the different minicircle sequence classes was estimated from the relative frequency of clones in the libraries and by quantitative Southern analysis of MspI-digested kDNA (Maslov and Simpson, 1992). The two major sequence classes (Lt19 and B4) are present in 3000 and 2500 copies per network of 10,000 minicircles and the remaining 15 sequence classes vary in copy number from 30 to 1000 molecules per network. There was no correlation of the steady-state gRNA abundance with specific minicircle DNA copy number, suggesting that gRNA abundance is determined by promotor strength or rates of degradation rather than through a gene-dosage effect.

In general, the presence of gRNA genes on low-copy-number minicircles, that apparently segregate randomly to daughter mitochondria, may prove to have significant genetic consequences in terms of the long-term stability of the editing system.

3.10 EDITING OF G1-G5 IN *LEISHMANIA*?

The transcript of the G1 region in *T. brucei* was shown to be productively edited to form an mRNA that could encode a protein (ND8) which has homology with a component of complex I of the mitochondrial electron-transport chain (Souza *et al.*, 1992; see Chapter 2). However, attempts to clone and sequence directly mature edited transcripts from G-rich regions 1 to 5 in *L. tarentolae* proved unsuccessful. Partially edited RNAs from G1 and G5 were detected, which appeared to represent several classes of unexpected editing patterns. From the sizes of the G-rich regions, we estimate that an additional 40-45 gRNAs would be required for complete editing of G1-G5 transcripts (Table 3). The fact that all 12 newly discovered minicircle-encoded gRNAs in *L. tarentolae* mediate editing of only the COIII, MURF4 and RPS12 (= G6) mRNAs makes it unlikely that the additional predicted gRNAs for G1-G5 actually exist in this strain. It is theoretically possible that these are maxicircle-encoded, but this is unlikely since all but one of the known gRNAs for pan-edited genes are minicircle-encoded in *L. tarentolae*. Furthermore, the informational part of the maxicircle is very limited in terms of intergenic regions which could encode additional gRNAs, and the presence of gRNA genes within the divergent region is unlikely since this consists of tandem repeats of AT-rich sequences, and few unique sequences. We speculate that the G1-G5 regions represent "pseudo-cryptogenes" in this strain of *L. tarentolae* because of a loss at mitochondrial division of low-copy-number minicircles encoding crucial gRNAs required for editing of these transcripts. The UC strain of *L. tarentolae* has been in culture for over 50 years. We suggest that the products of the G1-G5 cryptogenes may not be required for life in culture, which is usually equivalent to the insect phase of the life cycle. The finding that editing of the G1 mRNA in *T. brucei* is regulated and is limited to the bloodstream or animal phase of the life cycle (Souza *et al.*, 1992) is entirely consistent with this hypothesis. Also in line with this hypothesis is a previous observation that the kinetoplast minicircle DNAs from several recently isolated strains of *L. tarentolae* are uniformly more complex in terms of restriction patterns in gels than minicircle DNA from the UC strain, indicating larger minicircle-encoded gRNA repertoires in the former strains (Gomez-Eichelmann *et al.*, 1988).

3.11 THE ENZYME-CASCADE MODEL OF RNA EDITING

The presence of a 3' terminal uridylyl transferase (TUTase) in a purified kinetoplast-mitochondrial fraction from *L. tarentolae* was shown by Bakalara *et al.* (1989). This enzyme activity adds up to nine U residues to a 3' OH of an RNA molecule in an apparently non-specific fashion. Heparin was found to inhibit this activity (Simpson *et al.*, 1992). The mitochondrial gRNAs represent an especially good substrate for this activity. The evidence for a mitochondrial localization of the TUTase activity includes the organellar labeling of endogenous RNAs in the presence of the non-penetrating inhibitor, heparin, the lack of effect of added proteinase K on the activity in intact mitochondria, and the isopycnic co-banding of the activity with a mitochondrial marker enzyme in Percoll and Renogafin density gradients. A TUTase activity which adds a single U residue to the 3'-OH of RNAs has been described in whole-cell extracts of *T. brucei* (White and Borst, 1987), but the relationship between this activity and the mitochondrial activity is unknown. Moreover, an $[\alpha\text{-}^{32}\text{P}]$UTP incorporation activity specific for pre-edited messenger RNAs has been found in mitochondrial extracts of *T. brucei* (Harris *et al.*, 1990). This activity is independent of transcription, providing additional evidence for the post-transcriptional nature of RNA editing (Harris *et al.*, 1990).

An RNA ligase activity was also described in *L. tarentolae* by Bakalara *et al.* (1989). This activity required ATP and co-banded with the TUTase and a mitochondrial marker enzyme in isopycnic density gradients. A similar activity had been described previously for whole-cell extracts of *T. brucei* (White and Borst, 1987).

Inhibition of the mitochondrial TUTase activity by heparin or by digestion with proteinase K or pronase was found to activate a cryptic endonuclease which cleaved pre-edited CYb mRNA within the pre-edited region close to the first editing site (Simpson *et al.*, 1992). The cleavage activity was inhibited by SDS or phenol/chloroform extraction, and was retained by a 10 kd filter but passed through a 30 kd filter. The role of this nuclease is unknown. It could represent an RNA processing enzyme or, more likely, could play a role in RNA editing. A similar endonuclease activity was described by Harris *et al.* (1992) in mitochondrial extracts of *T. brucei*. This activity was found in extracts from mitochondria which were depleted of endogenous nucleotides by preincubation and the activity was stimulated by the addition of heparin. The endonuclease cleaved pre-edited but not edited mRNAs for cytochrome b, COII and COIII within editing domains. The major cleavage site for CYb mRNA was identical to that found in the case of *L. tarentolae*.

Based on the detection of TUTase, RNA ligase activities, and a site-specific cleavage activity within the mitochondrial-kinetoplast fraction of *L. tarentolae*, we originally proposed the enzyme-cascade model for the mechanism of gRNA-mediated RNA editing (Blum *et al.*, 1990) (Fig. 9). In this model, the initial interaction between the gRNA and the pre-edited mRNA is the formation of an anchor duplex just 3' of the pre-edited region. An endonuclease was invoked which recognizes the mismatch between the gRNA and the mRNA due to the presence of extra "guide" nucleotides in the gRNA and cleaves 3' of the mismatch. The TUTase would add a U to the released 3' hydroxyl of the mRNA. The two mRNA fragments would be joined by the RNA ligase, probably mediated by the complementarity of the added U residue with the guide A or G residue. This "mismatch-repair" process would proceed in a 5' direction as the gRNA-mRNA duplex is extended to the next mismatch. Deletions were postulated to be due to 3' exonuclease trimming of exposed and unpaired U residues at the cleavage site.

3.12 THE TRANSESTERIFICATION MODEL OF RNA EDITING

In the enzyme-cascade model for RNA editing, the 3' oligo-[U] tail of the gRNA played no role beyond stabilizing the initial hybrid by forming a duplex with the G+A-rich PER sequence (Blum and Simpson, 1990). A more active role for the 3' oligo-[U] tail was proposed in the transesterification model for RNA editing (Blum *et al.*, 1991; Cech, 1991) (Fig. 9). In this scenario, the initial event after formation of the gRNA-mRNA anchor duplex is a cleavage-ligation attack by the 3' OH of the gRNA at the mismatch, giving rise to a gRNA-mRNA chimeric

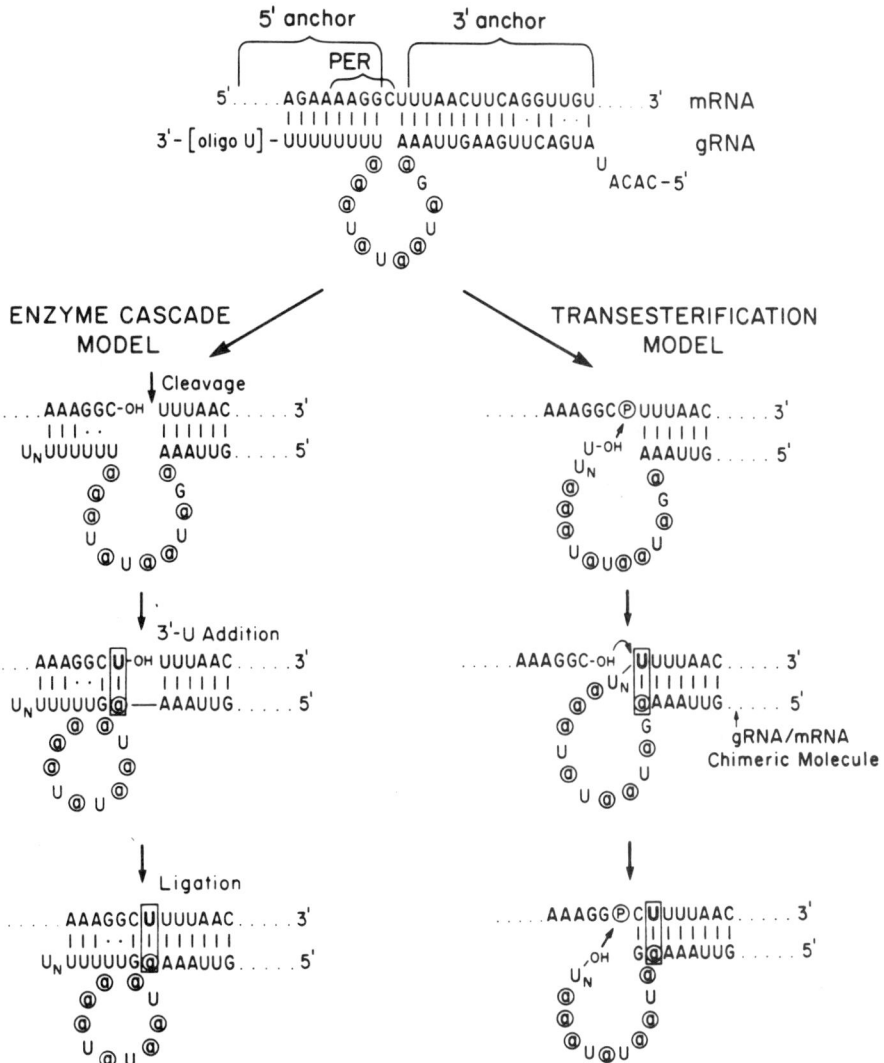

Fig. 9 — Two models for the mechanism of RNA editing in kinetoplastid mitochondria. PER, pre-edited region of CYb mRNA. The gRNA shown represents gCYb-I gRNA. Guide nucleotides are given in lower case and circled.

molecule. The 3' terminal U residues of the gRNA base pair with the internal guide sequence of the gRNA, thereby activating the first transesterification reaction in a manner similar to that occuring in the ribosomal RNA self-splicing intron of *Tetrahymena* (Davies *et al*., 1982; Waring and Davies, 1984). A second transesterification between the 3' OH of the cleaved mRNA fragment and the next mismatch liberates the gRNA and transfers at least one U residue into the mRNA at an editing site. The driving force towards incorporation of Us may stem from the elongation of the gRNA-mRNA duplex. This mechanism is formally very similar to the reversal of RNA splicing, which was observed *in vitro* using a group II self-splicing ribozyme (Moerl and Schmelzer, 1990a,b).

A prediction of this model is the existence of transient gRNA-mRNA chimeric molecules. These have been detected by selective PCR amplification from steady-state kRNA for four genes - ND7, COIII, COII and RPS12 (Blum *et al*., 1991; Maslov *et al*., 1992). *In vivo* chimerics almost always show the gRNA attached at normal editing sites, and downstream sites are fully and correctly edited. Similar chimerics have been detected in the case of *T. brucei*, in some cases in cDNA libraries constructed without PCR amplification (Read *et al*., 1992a; Koslowsky *et al*., 1991). It is possible, however, that chimeric molecules could be formed in terms of the enzyme-cascade model by adventitious ligation of the 3' end of the gRNA to the mRNA at the cleavage site. Therefore, the presence of chimerics does not prove the transesterification model but it is certainly consistent with this model.

In the transesterification model, the role of the TUTase is to add U residues to the 3' OH of the gRNA after transfer to the mRNA. The mitochondrial RNA ligase would have no obvious role in this model. In the enzyme-cascade model, the protease- or heparin-activated ribonuclease activity could represent the predicted endonuclease which initiates the editing process. Alternatively, in the transesterification model, this cleavage activity could actually represent the catalytic core for transesterification, which in this case catalyzes site-specific hydrolysis at normal transesterification sites in the absence of activated gRNAs. gRNAs, which are known to be bound to an 8S-19S complex of TUTase activity and several other proteins, are thought to be activated prior to utilization in editing (Bakker *et al*., submitted). Heparin could possibly prevent this activation by inhibiting the endogenous TUTase and releasing some low-affinity proteins from the complex, and protease digestion would destroy the complex, thereby releasing inactive gRNAs.

3.13 FORMATION OF gRNA/mRNA CHIMERIC MOLECULES *IN VITRO*

Two recent reports have shown the presence of an activity in mitochondrial extracts from *T. brucei* that promotes the formation of gRNA/mRNA chimeric molecules *in vitro*. Harris and Hajduk (1992) used labeled synthetic gRNA or mRNA to monitor the formation of chimeric molecules for the CYb cryptogene by gel electrophoresis, and Koslowski *et al*. (1992) used PCR amplification to analyze gRNA/mRNA chimeric products obtained using ATPase 6 (= MURF4) pre-edited mRNA and gA6-14 gRNA. The activity required a 3' OH on the gRNA substrate and appeared specific for pre-edited mRNA. We have obtained similar results for *L. tarentolae* (Blum and Simpson, 1992). Synthetic pre-edited messenger RNA and synthetic gRNA for the ND7 cryptogene from *L. tarentolae* formed chimeric molecules upon incubation in the presence of a mitochondrial extract. These chimeric molecules consisted of the gRNAs covalently linked to the mRNAs by short oligo-[U] tails at normal editing sites in most cases. Unlike the chimerics present *in vivo* in steady-state kinetoplast RNA, *in vitro* the chimeric molecules showed no editing of downstream editing sites. The formation of chimeric molecules required ATP and was dependent on the formation of a gRNA/mRNA anchor duplex 3' of the pre-edited region, as shown *in vitro* by mutagenesis of the mRNA and restoration of activity by compensatory base changes in the gRNA. mRNA sequences 3' and 5' of the pre-edited region also affected the efficiency of the chimeric molecule-forming activity.

The development of an accurate and complete editing system *in vitro* will be required to distinguish definitively between these models. The availability of an *in vitro* system for the formation of chimeric gRNA/mRNA molecules should allow a precise dissection of the sequence requirements for chimeric formation and also a fractionation of the extract components required for this activity.

3.14 RIBONUCLEOPROTEIN COMPLEXES INVOLVED IN RNA EDITING

Bakker *et al.* (submitted) showed that gRNAs in the kinetoplast mitochondrion of *L. tarentolae* are not free, but are bound to high-molecular-weight protein complexes which also contain TUTase activity. In glycerol gradients the TUTase-gRNA complexes vary in size from 8S to 19S in different mitochondrial preparations. The bound RNA can be labeled with $[\alpha-^{32}P]UTP$ by utilizing the endogenous TUTase activity. The chimeric-forming activity in preliminary experiments was shown to migrate at approximately 15-25S in glycerol gradients (Blum, Bakker and Simpson, unpublished results), depending on the isolation procedure. Incubation of labeled synthetic ND7 gRNA with mitochondrial extract resulted in the reconstitution of four major complexes, several of which comigrated with the native complexes. The close interaction of at least 6 proteins with the 3' oligo-[U] tails of the gRNAs in the complexes *in vivo* was shown by UV cross-linking experiments, using endogenously labeled gRNA (Bakker *et al.*, submitted).

It is likely that these ribonucleoprotein complexes represent part of the machinery involved in RNA editing.

3.15 PARTIALLY EDITED RNAs AND THE PROBLEM OF UNEXPECTED EDITING PATTERNS AT JUNCTION REGIONS

Editing intermediates which are edited only in the 3' portion of an editing domain can be isolated by PCR amplification of mitochondrial RNA, using a downstream primer for the cDNA synthesis which is specific for the edited sequence and an upstream primer for amplification specific for the pre-edited sequence. Partially edited CYb mRNA molecules from *L. tarentolae* displayed patterns of editing that were consistent with a strictly progressive 3' to 5' editing process. Only four out of 106 CYb clones showed unexpected editing patterns in which upstream editing preceded downstream editing or in which incorrect editing events occurred (Sturm and Simpson, 1990a). In the case of COIII, 177 out of 304 clones obtained exhibited strictly progressive 3' to 5' patterns of editing. The remaining 127 clones displayed unexpected editing patterns at the junction regions between fully edited and pre-edited sequences (Sturm and Simpson, 1990a). Unexpected editing patterns were also observed for the RPS12 gene in *L. tarentolae*. In a library of 26 different clones of partially edited RPS12 mRNAs, 8 clones showed unexpected editing patterns at junction regions (Maslov *et al.*, 1992). In the case of *T. brucei*, the majority of a library of partially edited mRNAs for CYb and COIII exhibited unexpected editing patterns at junction regions (Decker and Sollner-Webb, 1990).

We have proposed that unexpected editing patterns are generated by "misediting" due to specific events of "misguiding" (Sturm *et al.*, 1992). Misediting can occur by the interaction of inappropriate gRNAs with mRNAs or the appropriate gRNAs in an incorrect fashion. Four basic mechanisms by which misedited sequences could be generated by misguiding in a strict 3' to 5' fashion have been suggested on the basis of circumstantial evidence (Fig. 10). The formation of anchor duplexes by inappropriate gRNAs represents one type of initiating event in misediting. The formation of a secondary anchor, which is created independently or as a result of a loopout or bulge in either the gRNA or the mRNA, causes a shift in the gRNA/mRNA guiding frame and thereby misediting. This mechanism, which was first observed in a gRNA/mRNA chimeric by Blum *et al.* (1991), has been expanded in the progressive realignment model of Koslowsky *et al.* (1991) to include multiple misalignments and misediting occurring as normal intermediates of the editing process. These authors, therefore, refer to junction sequences as 'incompletely' edited; see Chapter 2. Frameshift misediting can also be produced by the mishybridization of a gRNA guiding G residue to a U in the mRNA which is supposed to be deleted. In this case the guiding G would normally hybridize to a C residue of the mRNA which is located just 5' of the U (or Us) to be deleted. Finally, some other misediting events occur at the very 3' end of the gRNAs, possibly reflecting gRNA 3' heterogeneity or misediting mediated by the oligo-[U] tail. Examples of each type of misediting have been observed, and misguiding events frozen in the form of chimeric RPS12 gRNA/COIII mRNA misedited molecules have been isolated by PCR amplification (Sturm *et al.*, 1992).

An alternative explanation for misediting within junction regions has also been proposed. Decker and Sollner-Webb (1990) suggested that editing occurs randomly at multiple sites within a defined region, with the hybridization of the gRNA to the correctly edited mRNA protecting the mRNA from further random editing.

The above-mentioned correlations of misedited patterns and gRNAs provide strong, though indirect, evidence in favor of the misguiding hypothesis. If unexpected patterns were derived from random editing, then the probability of

REGULAR EDITING

mRNA   pre-edited   edited   3'

5'   gRNA   Anchor

3'
U_n

MIS-EDITING by MIS-GUIDING

① mis-edited

"false" Anchor

U_n

②

"secondary" Anchor

U_n

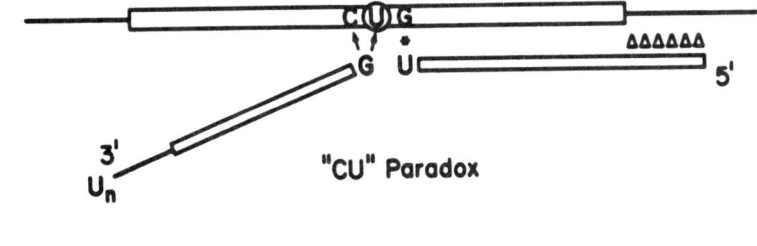

③ CUG

G   U

3'
U_n "CU" Paradox

④ U_n 5'

"internal" Anchor

U_n

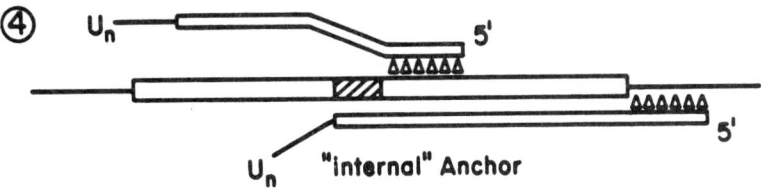

Fig. 10 — Schematic representation of normal editing and misediting produced by misguiding. Edited and pre-edited sequences of mRNA are shown as open boxes. Misedited sequences are indicated by cross-hatching and anchor sequences are represented by carets. Reprinted from Sturm *et al.* (1992) with permission.

finding any pattern correlations at all between unexpected patterns and gRNA sequences would be extremely low. It is, however, entirely possible that the editing mechanism in the case of *T. brucei* is less precise than in the case of *L. tarentolae*, possibly as a result of the greater diversity of gRNAs in the former.

The preferential localization of misediting to the junction region within an editing domain appears functional, since misediting outside an editing domain could not be repaired by re-editing. Limitation to a junction region is a natural consequence of the mechanisms proposed by Sturm *et al.* (1992), in which a ribonucleoprotein complex restricts editing to a certain defined region. In the case of false anchor formation (Fig. 10-1), the localization of misediting could be a result of the same mechanism which limits normal editing to a junction region. Bakker *et al.* (submitted) have suggested that the gRNA-TUTase complex is involved with the specific presentation of activated gRNAs to an editing site. Also, Harris and Hajduk (1992), Koslowsky *et al.* (1992) and Blum and Simpson (1992) have shown *in vitro* that the formation of gRNA/mRNA chimeric molecules requires the presence of proteins in the mitochondrial extract.

3.16 EVOLUTIONARY CONSIDERATIONS

The evolutionary origin of the kinetoplastid mitochondrial type of RNA editing is obscure. Since editing has been shown to exist in cells from the digenetic genera *Trypanosoma* and *Leishmania* (Simpson and Shaw, 1989), and also in species from the monogenetic genera *Crithidia* (Benne *et al.*, 1986; Van der Spek *et al.*, 1990, 1991) and *Herpetomonas* (Landweber and Gilbert, 1992), it is probably a primitive and not a derived character within the Trypanosomatidae family. It is an open question whether editing also occurs in kinetoplastids from the Bodonina suborder. It is known that an organized kinetoplast DNA network containing minicircles does not exist in cells from this taxonomic group (Hadjuk *et al.*, 1986; Maslov, D.A., Zajicek, P. and Benada, O., unpublished results), but the precise nature and coding capacity of the mitochondrial DNA present in these cells is unknown.

The flagellates, consisting of the kinetoplastids and the euglenoids, represent one of the earliest eukaryotic lineages, and may represent the first cell lineage with mitochondria (Sogin, 1989). It is possible that editing existed in an ancestor to the flagellates, but it is equally possible that editing evolved in parallel with the evolution of kinetoplastids. To answer this question, it is necessary to know whether editing occurs in Bodonid and in Euglenoid mitochondria. It would also be of interest to examine the possibility that editing occurs in an aerobic eubacterial descendent of the ancestral proto-mitochondrion which gave rise to the kinetoplast mitochondrion.

Comparative analysis of the editing patterns for homologous cryptogenes in *Leishmania* and *Trypanosoma* suggests that evolution of cryptogenes may have involved retroposition of partially edited RNAs into the maxicircle genome-replacing pan-edited genomic versions. For example, the ND7 mRNA in *T. brucei* is pan-edited apparently in two domains, whereas the ND7 mRNA in *L. tarentolae* (and *C. fasciculata*) is internally edited precisely at the 5' terminus of domain I and 5'-edited precisely at the 5' terminus of domain II (see Fig. 2). There is a similar situation in the case of the MURF4 mRNA and also the COIII mRNA. The absence of a poly-[A] sequence at the 3' terminus of the putative retroposed genomic copies is not unprecedented in other transposable systems. This hypothesis would suggest that pan-editing is the ancient type of editing, and that the internally edited and 5'-edited cryptogenes, and possibly also the completely unedited genes, were derived from originally pan-edited cryptogenes. In this scenario, the original multiple gRNA genes for the pan-edited RNAs would be lost in the cell containing the retroposed cryptogene because of lack of selection. In fact, the loss of low-copy-number minicircles by random segregation at network division could create a positive selection for the retroposition of partially edited RNAs back into the genome and possibly also for the transfer of gRNA genes from minicircles to maxicircles.

Retroposition of fully or partially edited mRNAs into the maxicircle genome would successively lead to the elimination of pan-edited genes. However, duplication of a small part of a mitochondrial gene into a transcription unit which activates transcription in the opposite direction could create a new potential gRNA. As soon as T deletions or insertions occur in the mRNA gene, the gRNA would become functional, and both sequences would then be able to drift away from the original sequence.

The cryptogene sequence must co-evolve in a very specific way with the gRNA sequence to maintain the proper editing process. Any U insertions or deletions within the portion of an mRNA gene which is covered by a gRNA will be corrected by editing. In regard to the gRNA genes, substitution of a guiding A with a guiding G or vice versa is permissive.

One possible selective pressure for the evolution of G-U base pairs in the gRNA/mRNA hybrid could be the necessity to melt the downstream hybrid in order to form a stable anchor duplex with the upstream gRNA to continue the editing. Moreover, the genes for the gRNAs and mRNAs can, in principle, substitute any C for a U residue if paired with a G residue, although G-U base pairs are rare in the anchor portion of the gRNA-mRNA hybrid, most probably because of lack of stability. The even weaker A-C base pair occurs very infrequently in gRNA/mRNA hybrids (Van der Spek *et al.*, 1991) and is always asymmetric, with A being present in the mRNA and C in the gRNA; the occurrence of A in the gRNA would lead to an addition of U in the mRNA in terms of the editing model. This suggests that even silent T-to-C transitions in the mRNA sequence may have a profound effect, leading to changes in the "guiding frame" similar to those observed in the misediting process and thereby resulting in disruption of the editing process within a domain. The T-to-C transitions can also destroy the editing process in the case in which the T corresponded to a U which is normally deleted. C-to-T transitions do not have such profound consequences for editing. This evolutionary pressure may explain in part the observed T-richness of the coding strand of maxicircle genes.

3.17 THE RELATIONSHIP OF RNA EDITING TO RNA SPLICING
The presence of gRNA/mRNA chimeric molecules in steady-state kinetoplastid

mitochondrial RNA raised the possibility that the mechanism of RNA editing might proceed through successive transesterification reactions (Blum *et al.*, 1991). In this model, uridine residues are transferred from the oligo-[U] tail of the guide RNA into the messenger RNA in a reaction which can be considered as a true reversal of the RNA-splicing mechanism. The inserted uridines can be regarded as the smallest introns known (Cech, 1991). It is noteworthy that both in Group I and in Group II self-splicing, the excision of the intron can be precisely reversed *in vitro* (Moerl and Schmelzer, 1990a; Doudna and Szostak, 1989; Sullivan and Cech, 1985). It is the prospect of a direct relationship between RNA editing and RNA splicing which makes the transesterification model for RNA editing so attractive.

In most RNA-catalyzed processes such as RNA splicing, RNase P cleavage, and possibly the peptidyl transferase activity of the ribosomal RNA (Noller *et al.*, 1992), proteins are believed to stabilize intra- and intermolecular RNA structures, thereby decreasing the requirements for a fixed primary sequence. This may explain the relatively low sequence conservation observed in nuclear cis-splicing, in which proteins and ribonucleoprotein particles appear to have assumed some of the structural functions probably originally accomplished by the RNA intron itself. In the case of RNA editing, which necessarily lacks any conserved sequence dependence at all, the participation of a ribonucleoprotein complex may confer the required specificity in addition to base-pairing interactions at the anchor sequence.

Assuming that RNA editing and RNA splicing share a common chemistry, we can speculate about the evolutionary origin of both processes. The basic result of a transesterification reaction is the exchange of partner RNA molecules, which allows RNA sequences to be precisely rearranged, given that they are in the proper configuration to activate a transfer. In this view, introns and gRNAs may both represent relics of the same early genetic systems which dealt with information management of RNA molecules. RNA splicing (including trans-splicing) and its reversal would have allowed for relatively large sequences to be rearranged, a process which is referred to as "RNA recombination" (Moerl and Schmelzer, 1990b). RNA editing, on the other hand, could have been used to change locally an RNA sequence.

In the modern kinetoplast mitochondrion, only the U content of mRNAs can be revised, a process which is directly reflected by the presence of the unique oligo-U 3' tail of the gRNAs. It is possible that other nucleotides than Us were rearranged as well in an ancestral world by separate gRNAs containing a different oligo-homopolymer 3'-tail, specific for the nucleotide insertion. It is possibly relevant that a specific C-insertion mechanism has been reported in mitochondria of *Physarum polycephalum* (Mahendran *et al.*, 1991, see Chapter 4) and it will be interesting to see whether gRNAs are involved in this form of RNA editing. It is conceivable that any RNA sequence can be changed locally into any other RNA sequence given appropriate gRNAs and enzymatic machinery. An ancestral genetic system possessing RNA splicing and RNA editing machineries would have the potential to create new diversity using existing RNA oligonucleotides.

A problem which possibly confronted the "RNA world" was the limited stability of RNA molecules capable of catalytic activities (Joyce, 1989). One solution to this problem could have been to separate the RNA population into an early "genotype" and "phenotype". The genotype would consist of short RNA molecules, which would be relatively stable. The phenotype, on the other hand, would consist of the larger catalytically active molecules which could be assembled from the smaller molecules in a series of guided transesterifications. After the transition to the DNA world, the information for one gene may have continued to remain split up into several small pieces at different locations in the rudimentary DNA genome, requiring both RNA splicing and RNA editing to construct complete mRNAs for translation. In this scenario, modern RNA editing in kinetoplastid protozoa represents a relic of a primitive mechanism for creating functional mRNA sequences.

REFERENCES

Bakalara, N., Simpson, A.M. & Simpson, L. (1989). The *Leishmania* kinetoplast-mitochondrion contains terminal uridylyltransferase and RNA ligase activities. *J. Biol. Chem.* **264** 18679-18686.

Benne, R., Van den Burg, J., Brakenhoff, J., Sloof, P., Van Boom, J. & Tromp, M. (1986). Major transcript of the frameshifted coxII gene from trypanosome mitochondria contains four nucleotides that are not encoded in the DNA. *Cell* **46** 819-826.

Bhat, G.J., Koslowsky, D.J., Feagin, J.E., Smiley, B.L. & Stuart, K. (1990). An extensively edited mitochondrial transcript in kinetoplastids encodes a protein homologous to ATPase subunit 6. *Cell* **61** 885-894.

Bhat, G.J., Myler, P.J. & Stuart, K. (1991). The two ATPase 6 mRNAs of *Leishmania* tarentolae differ at their 3' ends. *Mol. Biochem. Parasitol.* **48** 139-150.

Blum, B., Bakalara, N. & Simpson, L. (1990). A model for RNA editing in kinetoplastid mitochondria: "guide" RNA molecules transcribed from maxicircle DNA provide the edited information. *Cell* **60** 189-198.

Blum, B., Sturm, N.R., Simpson, A.M. & Simpson, L. (1991). Chimeric gRNA-mRNA molecules with oligo(U) tails covalently linked at sites of RNA editing suggest that U addition occurs by transesterification. *Cell* **65** 543-550.

Blum, B. & Simpson, L. (1990). Guide RNAs in kinetoplastid mitochondria have a nonencoded 3' oligo-(U) tail involved in recognition of the pre-edited region. *Cell* **62** 391-397.

Blum, B. & Simpson, L. (1992) Formation of gRNA/mRNA chimeric molecules *in vitro*, the initial step of RNA editing, is dependent on an anchor sequence. *Proc. Natl. Acad. Sci. USA* **89** 11944-11948.

Cech, T.R. (1991). RNA editing: world's smallest introns. *Cell* **64** 667-669.

Davies, R., Waring, R., Ray, J., Brown, T. & Scazzocchio, C. (1982). Making ends meet: a model for RNA splicing in fungal mitochondria. *Nature* **300** 719-724.

De la Cruz, V., Neckelmann, N. & Simpson, L. (1984). Sequences of six structural genes and several open reading frames in the kinetoplast maxicircle DNA of *Leishmania* tarentolae. *J. Biol. Chem.* **259** 15136-15147.

Decker, C.J. & Sollner-Webb, B. (1990). RNA editing involves indiscriminate U changes throughout precisely defined editing domains. *Cell* **61** 1001-1011.

Doudna, J.A. & Szostak, J.W. (1989). RNA-catalyzed synthesis of complementary strand RNA. *Nature* **339** 519-522.

Feagin, J.E., Abraham, J. & Stuart, K. (1988a). Extensive editing of the cytochrome c oxidase III transcript in *Trypanosoma brucei*. *Cell* **53** 413-422.

Feagin, J.E., Shaw, J.M., Simpson, L. & Stuart, K. (1988b). Creation of AUG initiation codons by addition of uridines within cytochrome b transcripts of kinetoplastids. *Proc. Natl. Acad. Sci. USA* **85** 539-543.

Gomez-Eichelmann, C., Holz, G., Beach, D., Simpson, A. & Simpson, L. (1988). Comparison of several lizard *Leishmania* species and strains in terms of kinetoplast minicircle and maxicircle DNA sequences, nuclear chromosomes and membrane lipids. *Mol. Biochem. Parasitol.* **27** 143-158.

Gribskov, M., Luthy, R. & Eisenberg, D. (1989). Profile analysis. In Methods in Enzymology, "Molecular Evolution: Computer Analysis of Protein and Nucleic Acid Sequences." R. F. Doolittle, ed., Academic Press, vol. 183.

Hajduk, S.L., Siqueira, A.M. & Vickerman, K. (1986). Kinetoplast DNA of Bodo caudatus: a noncatenated structure. *Mol. Cell Biol.* **6** 4372-4378.

Harris, M., Decker, C., Sollner-Webb, B. & Hajduk, S. (1992). Specific cleavage of pre-edited mRNAs in trypanosome mitochondrial extracts. *Mol. Cell. Biol.* **12** 2591-2598.

Harris, M.E., Moore, D.R. & Hajduk, S.L. (1990). Addition of uridines to edited RNAs in trypanosome mitochondria occurs independently of transcription. *J. Biol. Chem.* **265** 11368-11376.

Harris, M.E. & Hajduk, S.L. (1992). Kinetoplastid RNA editing: *in vitro* formation of cytochrome b gRNA-mRNA chimeras from synthetic substrate RNAs. *Cell* **68** 1091-1099.

Hensgens, A., Brakenhoff, J., de Vries, B., Sloof, P., Tromp, M., Van Boom, J. & Benne, R. (1984). The sequence of the gene for cytochrome c oxidase subunit I, a frameshift containing gene for cytochrome c oxidase subunit II and seven unassigned reading frames in *Trypanosoma brucei* mitochondrial maxi-circle DNA. *Nucl. Acids Res.* **12** 7327-7344.

Joyce, G. (1989). RNA evolution and the origins of life. *Nature* **338** 217-224.

Koslowsky, D.J., Bhat, G.J., Perrollaz, A.L., Feagin, J.E. & Stuart, K. (1990). The MURF3 gene of *T. brucei* contains multiple domains of extensive editing and is homologous to a subunit of NADH dehydrogenase. *Cell* **62** 901-911.

Koslowsky, D.J., Bhat, G.J., Read, L.K. & Stuart, K. (1991). Cycles of progressive realignment of gRNA with mRNA in RNA editing. *Cell* **67** 537-546.

Koslowsky, D.J., Göringer, H.U., Morales, T.H. & Stuart, K. (1992). *In vitro* guide RNA/mRNA chimaera formation in *Trypanosoma brucei* RNA editing. *Nature* **356** 807-809.

Landweber, L. & Gilbert, W. (1992). Extensive editing of the cytochrome c oxidase III transcript in the *Herpetomonas* genus: the evolution of RNA editing. *J. Cell. Biochem.* Suppl. **16A**, 149.

Mahendran, R., Spottswood, M.R. & Miller, D.L. (1991). RNA editing by cytidine insertion in mitochondria of *Physarum polycephalum*. *Nature* **349** 434-438.

Maslov, D.A., Sturm, N.R., Niner, B.M., Gruszynski, E.S., Peris, M. & Simpson, L. (1992). An intergenic G-rich region in *Leishmania tarentolae* kinetoplast maxicircle DNA is a pan-edited cryptogene encoding ribosomal protein S12. *Mol. Cell. Biol.* **12** 56-67.

Maslov, D.A. & Simpson, L. (1992). The polarity of editing within a multiple gRNA-mediated domain is due to formation of anchors for upstream gRNAs by downstream editing. *Cell* **70** 459-467.

Moerl, M. & Schmelzer, C. (1990a). Integration of group II intron bI1 into a foreign RNA by reversal of the self-splicing reaction *in vitro*. *Cell* **60** 629-636.

Moerl, M. & Schmelzer, C. (1990b). Group II intron RNA-catalyzed recombination of RNA *in vitro*. *Nucleic Acids Res.* **18** 6545-6551.

Noller, H.F., Hoffarth, V. & Zimniak, L. (1992). Unusual resistance of peptidyl transferase to protein extraction procedures. *Science* **256** 1416-1419.

Pause, A. & Sonenberg, N. (1992). Mutational analysis of a DEAD box RNA helicase: the mammalian translation initiation factor eIF-4A. *EMBO J.* **11** 2643-2654.

Payne, M., Rothwell, V., Jasmer, D., Feagin, J. & Stuart, K. (1985). Identification of mitochondrial genes in *Trypanosoma brucei* and homology to cytochrome c oxidase II in two different reading frames. *Mol. Biochem. Parasitol.* **15** 159-170.

Pollard, V.W., Rohrer, S.P., Michelotti, E.F., Hancock, K. & Hajduk, S.L. (1990). Organization of minicircle genes for guide RNAs in *Trypanosoma brucei*. *Cell* **63** 783-790.

Pollard, V.W. & Hajduk, S.L. (1991). *Trypanosoma equiperdum* minicircles encode three distinct primary transcripts which exhibit guide RNA characteristics. *Mol. Cell. Biol.* **11** 1668-1675.

Read, L.K., Corell, R.A. & Stuart, K. (1992a). Chimeric and truncated RNAs in *Trypanosoma brucei* suggest transesterifications at non-consecutive sites during RNA editing. *Nucleic Acids Res.* **20** 2341-2347.

Read, L.K., Myler, P.J. & Stuart, K. (1992b). Extensive editing of both processed and preprocessed maxicircle CR6 transcripts in *Trypanosoma brucei*. *J. Biol. Chem.* **267** 1123-1128.

Shaw, J., Feagin, J.E., Stuart, K. & Simpson, L. (1988). Editing of mitochondrial mRNAs by uridine addition and deletion generates conserved amino acid sequences and AUG initiation codons. *Cell* **53** 401-411.

Shaw, J., Campbell, D. & Simpson, L. (1989). Internal frameshifts within the mitochondrial genes for cytochrome oxidase subunit II and maxicircle unidentified reading frame 3 in *Leishmania tarentolae* are corrected by RNA editing: evidence for translation of the edited cytochrome oxidase subunit II mRNA. *Proc. Natl. Acad. Sci. USA* **86** 6220-6224.

Simpson, A.M., Suyama, Y., Dewes, H., Campbell, D. & Simpson, L. (1989). Kinetoplastid mitochondria contain functional tRNAs which are encoded in nuclear DNA and also small minicircle and maxicircle transcripts of unknown function. *Nucl. Acids Res.* **17** 5427-5445.

Simpson, A.M., Bakalara, N. & Simpson, L. (1992). A ribonuclease activity is activated by heparin or by digestion with proteinase K in mitochondrial extracts of *Leishmania tarentolae*. *J. Biol. Chem.* **267** 6782-6788.

Simpson, L. (1972). The kinetoplast of the hemoflagellates. *Int. Rev. Cytol.* **32** 139-207.

Simpson, L. (1987). The mitochondrial genome of kinetoplastid protozoa: Genomic organization, transcription, replication and evolution. *Ann. Rev. Microbiol.* **41** 363-382.

Simpson, L., Neckelmann, N., de la Cruz, V., Simpson, A., Feagin, J., Jasmer, D. & Stuart, K. (1987). Comparison of the maxicircle (mitochondrial) genomes of *Leishmania* tarentolae and *Trypanosoma brucei* at the level of nucleotide sequence. *J. Biol. Chem.* **262** 6182-6196.

Simpson, L. & Shaw, J. (1989). RNA editing and the mitochondrial cryptogenes of kinetoplastid protozoa. *Cell* **57** 355-366.

Sogin, M. (1989). Phylogenetic meaning of the kingdon concept: an unusual ribosomal RNA from *Giardia lamblia*. *Science* **243** 75-77.

Souza, A.E., Myler, P.J. & Stuart, K. (1992). Maxicircle CR1 transcripts of *Trypanosoma brucei* are edited, developmentally regulated, and encode a putative iron-sulfur protein homologous to an NADH dehydrogenase subunit. *Mol. Cell. Biol.* **12** 2100-2107.

Sturm, N.R., Maslov, D.A., Blum, B. & Simpson, L. (1992). Generation of unexpected editing patterns in *Leishmania tarentolae* mitochondrial mRNAs: misediting produced by misguiding. *Cell* **70** 469-476.

Sturm, N.R. & Simpson, L. (1990a). Partially edited mRNAs for cytochrome b and subunit III of cytochrome oxidase from *Leishmania tarentolae* mitochondria: RNA editing intermediates. *Cell* **61** 871-878.

Sturm, N.R. & Simpson, L. (1990b). Kinetoplast DNA minicircles encode guide RNAs for editing of cytochrome oxidase subunit III mRNA. *Cell* **61** 879-884.

Sturm, N.R. & Simpson, L. (1991c). *Leishmania tarentolae* minicircles of different sequence classes encode single guide RNAs located within the variable region approximately 150 bp from the conserved region. *Nucl. Acids Res.* **19** 6277-6281.

Sullivan, F.X. & Cech, T. (1985). Reversibility of cyclization of the *Tetrahymena* rRNA intervening sequence: implication for the mechanism of splice site choice. *Cell* **42** 639-648.

Van der Spek, H., Van den Burg, J., Croiset, A., Van den Broek, M., Sloof, P. & Benne, R. (1988). Transcripts from the frameshifted MURF3 gene from *Crithidia fasciculata* are edited by U insertion at multiple sites. *EMBO J.* **7** 2509-2514.

Van der Spek, H., Speijer, D., Arts, G., Van den Burg, J., Van Steeg, H., Sloof,

P. & Benne, R. (1990). RNA editing in transcripts of the mitochondrial genes of the insect trypanosome, *Crithidia fasciculata. EMBO J.* **9** 257-262.

Van der Spek, H., Arts, G.-J., Zwaal, R.R., Van den Burg, J., Sloof, P. & Benne, R. (1991). Conserved genes encode guide RNAs in mitochondria of *Crithidia fasciculata. EMBO J.* **10** 1217-1224.

Waring, R.B. and Davies, R.W. (1984). Assessment of a model for intron RNA secondary structure relevant to RNA self-splicing: a review. *Gene.* **28** 277-291.

White, T. and Borst, P. (1987). RNA end-labeling and RNA ligase activities can produce a circular ribosomal RNA in whole cell extracts from trypanosomes. *Nucl. Acids Res.* **15** 3275-3290.

APPENDIX: TERMINOLOGY OF RNA EDITING

Chimeric gRNA/mRNA
molecules   - RNAs which consist of gRNA molecules covalently
                              linked by the oligo-[U] tail to mRNA at an editing
                              site.

Cryptogene   - gene whose transcript is edited.

Domain connection
sequence   - unedited sequence separating two editing domains
                              which provides an anchor sequence for the initial
                              gRNA.

Editing block   - the editing mediated by a single gRNA.

Editing domain   - a region of the mRNA in which editing occurs
                              independently of editing in other regions. Usually
                              involves multiple overlapping gRNAs.

Editing site   - site in mature edited RNA where Us are inserted or
                              deleted.

Enzyme-cascade model   - editing occurs by cleavage, U-addition or U-removal,
                              and ligation, mediated by base-pairing with gRNA.

G-rich   - corresponds to the guanine-versus-cytosine strand-
                              biased regions called CR in Chapter 2. These
                              regions are transcribed into G-rich RNAs, all of
                              which are extensively edited in *T. brucei*. In the *L.
                              tarentolae* UC strain, extensive editing has only
                              been demonstrated in G6 RNA.

Guide RNA (gRNA)   - small RNA complementary to mature edited RNA;
                              has non-encoded 3' oligo-[U] tail.

Guiding frame   - the reading frame in which the gRNA sequence will
                              correctly encode the U insertions and deletions.

Junction region   - the region of partially edited mRNA in an editing
                              domain between fully edited and pre-edited
                              sequences.

Kinetoplast   - the region of the single mitochondrion containing the
                              kDNA.

Kinetoplast nucleoid
body   - the kDNA network *in situ*.

Mature edited RNA   - the correctly (fully) edited RNA species.

Misediting   - unexpected editing pattern produced by misguiding
                              gRNAs.

Misguiding   - formation of incorrect anchor hybrid and subsequent
                              misediting.

Pan-editing   - RNA is edited over its entire length; may have
                              several editing domains.

Partially edited RNAs   - RNA molecules which are edited in the 3' region and
                              not in the 5' region of an editing domain.

Pre-edited region (PER)   - the region of the mRNA that becomes edited in the mature RNA, as distinguished from the unedited region.

Pseudo-cryptogene

- a cryptogene whose transcript is not productively edited to yield a mature mRNA because of the loss of one or more gRNA genes.

Transesterification model

- editing occurs by two succesive transesterifications transferring Us from gRNA oligo-[U] tail to mRNA, mediated by base-pairing with gRNA. Deletions occur either by U-specific exonuclease trimming of 3' end of mRNA or by transfer of U from mRNA to gRNA 3' tail.

Unexpected editing patterns

- editing patterns that show 5' editing before 3' editing is complete, or any pattern which differs from the fully edited pattern.

4

RNA EDITING IN MITOCHONDRIA OF *PHYSARUM POLYCEPHALUM*

D. Miller, R. Mahendran, M. Spottswood, M. Ling, S. Wang, N. Yang and H. Costandy
The Molecular and Cell Biology Program, The University of Texas at Dallas Richardson, Texas 75080, USA

RNAs synthesized in mitochondria of the acellular slime mold *Physarum polycephalum* require RNA editing to create open reading frames in mRNA and to generate functional rRNAs and tRNAs. The RNA editing in *Physarum* is distinct from other types of RNA processing which are also termed RNA editing. It is the only type of extensive insertional editing outside of the kinetoplastid protozoa but it is only superficially similar to the editing in those organisms. The insertional editing observed in *Physarum* is generally the addition of a single cytidine. However, the other three nucleotides are also occasionally inserted. Therefore, *Physarum* provides the only current example of mixed nucleotide insertional editing. Also, the substrates for the editing are unique in that in addition to mRNAs, tRNAs and rRNAs are also edited. Our studies indicate that most RNAs in *Physarum* mitochondria are edited and that editing is a fundamental and necessary step in mitochondrial gene expression.

4.1 *PHYSARUM POLYCEPHALUM*

Physarum polycephalum is a member of the myxomycota or plasmodial slime molds (Margulis and Schwartz, 1988). The members of this phylum are free-living and are characterized by a complex life cycle that includes both a fungal-like diploidstage and an amoeboid haploid stage. As a diploid *Physarum* grows as a plasmodium, i.e. nuclei are not separated by plasma membranes. Plasmodia are usually pigmented (yellow or orange) but do not photosynthesize. In the wild,

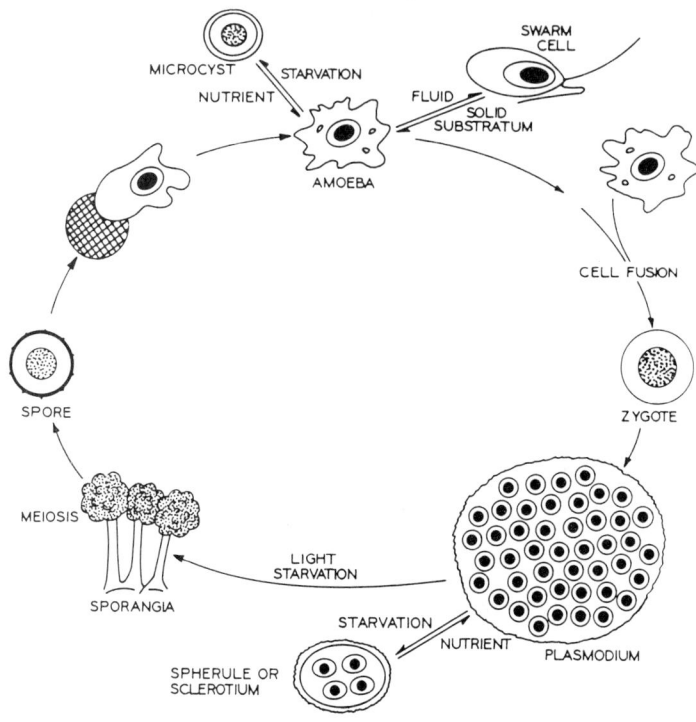

Fig. 1 —— Life cycle of *Physarum polycephalum*.

plasmodia feed on decaying vegetation and a semi-defined medium has been developed for growing plasmodia in the laboratory (Daniel and Baldwin, 1964). *Physarum* can be grown either as macroplasmodia on agar gels or on filters in petri dishes, or as microplasmodial suspensions in shaking liquid cultures. Desiccation, starvation or exposure to light induce plasmodia to grow stalked sporangia (fruiting bodies) in which meiosis takes place. Haploid myxamoeba are produced by germination of the spores. These myxamoeba can readily convert into undulipodiated (= having flagellae) cells called swarm cells. Myxamoeba or swarm cells of different mating types can fuse to form a zygote which undergoes repeated mitoses to reproduce the plasmodium (Fig. 1).

Physarum polycephalum has been a favorite experimental organism for studies of the cell cycle since in the plasmodium all of the thousands of nuclei are synchronized at the same point of the cell cycle. Another attraction of *Physarum* is the complex life cycle which provides a model of differentiation in an organism that can be readily grown in large cultures. We were initially interested in *Physarum* because of a third feature, i.e. its ancient divergence from the other eukaryotic kingdoms of plants, animals and fungi (Hasegawa *et al.*, 1985; Johansen *et al.*, 1988). It was our expectation that because of this ancient divergence, we would find unique mechanisms of mitochondrial gene expression and this has indeed been the case.

4.2 MITOCHONDRIAL DNA OF *PHYSARUM POLYCEPHALUM*

Mitochondrial (mt) DNA makes up 5-10% of the total plasmodial DNA. It is easily separated from the nuclear DNA in CsCl density gradients since its base composition is 75% A+T as opposed to the nuclear DNA which is about 60% A+T (Evans and Suskind, 1971). Digestion of the isolated mtDNA with restriction endonucleases indicates that the complexity of the mtDNA of *Physarum polycephalum* is about 60 kb (Jones *et al.*, 1990; Fig. 2). Mapping with restriction enzymes gives a circular map. We have succeeded in cloning mtDNA fragments representative of the entire mtDNA. These fragments have been used to produce a transcription map of the mtDNA, which indicates that about 75% of the mtDNA is heavily transcribed whereas about 25% is either untranscribed or transcribed at a very low level under the conditions studied. The transcription pattern as well as hybridization studies using radiolabeled fragments of corn mtDNA containing the genes for the large or small subunit rRNA have enabled us to identify the rRNA genes on *Physarum* mtDNA. The identity of these genes has been confirmed by sequence analysis (Mahendran *et al.*, 1992; Yang *et al.*, 1992). Concomitant with the sequence analysis of the rRNAs a number of tRNA genes were identified by computer using an algorithm which detects conserved primary sequences of tRNAs and the potential for folding into the classic cloverleaf structure. That these sequences actually encode tRNAs was confirmed by hybridization studies which showed that they hybridized to 4S RNA (Costandy *et al.*, 1992).

Once the genes for the large and small subunit rRNA had been identified it was assumed that the other major high molecular weight transcripts would be mRNAs. We planned to identify mRNA-coding regions by sequencing the transcribed regions of mtDNA and searching for open reading frames. We were initially successful in doing this and identified an open reading frame 411 codons in length (ORF1, see Fig. 2, Mahendran, 1990). This reading frame was considered highly significant since the average distance between termination codons in the other five transcribed regions was about 13 codons. Analysis of ORF1 revealed that codons rich in A or U were favored and that all three of the classic termination codons were excluded, indicating that UGA is a termination codon and does not code for tryptophan as in some mitochondria. This observation also holds for edited RNAs; see below. The identity of the protein inferred from the open reading frame could not be determined.

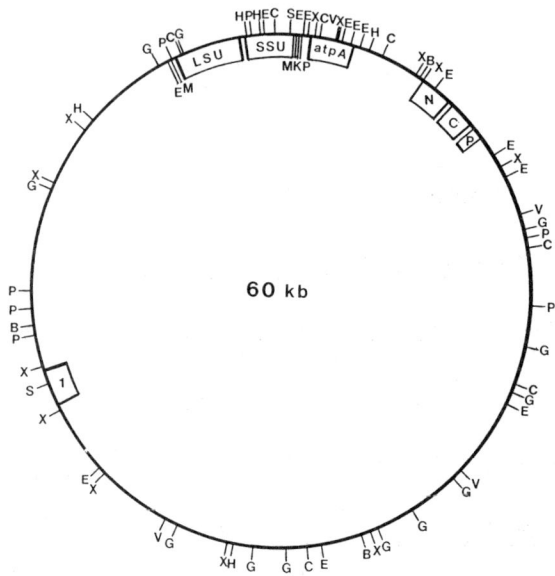

Fig. 2 — Mitochondrial DNA of *Physarum polycephalum* strain M3. The restriction enzyme map of the mtDNA forms a 60 kb circle (Jones *et al.*, 1990). Recognition sites of restriction enzymes BamHI (B), ClaI (C), EcoRI (E), PstI (P), SalI (S), XbaI (X), BglII (G), PvuII (V), and XhoI (H) are shown on the outside of the circle. Cryptogenes for the large subunit rRNA (LSU), small subunit rRNA (SSU), alpha subunit of ATP synthase (atpA), subunit 1 of the NADH dehydrogenase complex (N), apocytochrome b (C), and the proteolipid subunit of ATP synthase (P) as well as an unassigned open reading frame, ORF 1 (1), are shown as rectangles within the circle. Lines indicating tRNA genes are labeled with the one letter abbreviation of their cognate amino acid as deduced from the apparent anticodon.

4.3 DISCOVERY OF RNA EDITING IN *PHYSARUM POLYCEPHALUM*

Further sequence analysis of the mtDNA revealed that additional significant open reading frames were lacking in a heavily transcribed 7 kb region downstream of the rRNA genes. In order to identify protein genes in this region, a cDNA library made from total *Physarum* RNA was screened for homology to mtDNA probes. A 1.0 kb cDNA with homology to mtDNA sequences just downstream of the small subunit rRNA gene was identified (Mahendran *et al.*, 1991). Digestion of the cDNA with restriction enzymes revealed that while a number of restriction fragments corresponded to restriction fragments derived from the mtDNA, some unique restriction fragments were present. These unique fragments were due to the absence in the cDNA of some restriction sites found on the mtDNA and to the presence of additional restriction sites not found on the mtDNA. Sequence analysis of the cloned cDNA revealed a significant open reading frame. Assignment of amino acids using the classic genetic code produced a protein sequence with homology to alpha subunits of ATP synthase. The analogous region of the mtDNA lacked this reading frame although it contained extensive

```
                                                      *                    *            120
ATGACATTTCTTAAATCCGTATATATTTTAGATGATGAAATTAATCAAGATGTTATTGATAACATTGATGAAAATGAAGACCCGTCAATCGGTCAAATTATTTCTGTCAAAGATGGTGTT
M  T  F  L  K  S  V  Y  I  L  D  D  E  I  N  Q  D  V  I  D  N  I  D  E  N  E  D  P  S  I  G  Q  I  I  S  V  K  D  G  V

      *                        *                   *                *                *           240
GCTTTTGTTACAGGACTTGATAATATTCAAGTAGGTGAAATGGTAGAATTTATCTCTAAGGGTTTAACCGGTATGGCTCTTAATCTAGAAGCTGAACAAGTCGGTTGTATTATTTTTGGT
A  F  V  T  G  L  D  N  I  Q  V  G  E  M  V  E  F  I  S  K  G  L  T  G  M  A  L  N  L  E  A  E  Q  V  G  C  I  I  F  G

      *              *              *             *                  *                   *           360
GATGATACCTCTGTTCAACAAGATGATTCTGTCCGTGCTTTAAATACCTTAGTCAAAACCCCTGTAGGTTATGGCTTATTAGGTCGTGTTGTCGATGGTATTGGTAATTTCATCGATGGT
D  D  T  S  V  Q  Q  D  D  S  V  R  A  L  N  T  L  V  K  T  P  V  G  Y  G  L  L  G  R  V  V  D  G  I  G  N  F  I  D  G

      *              *                  *                     *                  *             360
GGTGAAACTATCGCTTTTGAGGAATATCTCAATGTCGAACGTAAAGCTCCTGGTGTTATCACTCGTGAATCTGTTACTGAACCAATGTTAACTGGTTATAAATCGTTGATTCTATGTTA
G  E  T  I  A  F  E  E  Y  L  N  V  E  R  K  A  P  G  V  I  T  R  E  S  V  T  E  P  M  L  T  G  Y  K  I  V  D  S  M  L

     *                 *               *                    *                    *            *600
CCTATCGGACGTGGTCAAAGGGAACTTATTGTTGGTGATCGTCAAACAGGTAAAACTACTATTGCTATCGATACTATTCTTAATCAACGTTATACCAATGAAGAAGATATTGATCTCTAT
P  I  G  R  G  Q  R  E  L  I  V  G  D  R  Q  T  G  K  T  T  I  A  I  D  T  I  L  N  Q  R  Y  T  N  E  E  D  I  D  L  Y

      *                 *              *              *                  *                   *           720
TGTGTGTATGTTGGAATTGGTCAAAAAAAAAGTTCTATCTTTGAATATTCAAACCTTACTTCAAAATAATAAAGCCGCTCATTATACTACTATTGTTGCTGCAACCGCTGCACAAAGTGCC
C  V  Y  V  G  I  G  Q  K  K  S  S  I  L  N  I  Q  T  L  L  Q  N  N  K  A  A  H  Y  T  T  I  V  A  A  T  A  A  Q  S  A

     *               *                *              *                        *            840
TCTCTTCAGTTTATTGCTCCATATACAGGATGTGCCATTGCTGAATTTTATCGTGATCAAGGTGAACATGCCTTAATTATTTATGATGATTTAACAAAACATGCTGCTGCTTATCGACAA
S  L  Q  F  I  A  P  Y  T  G  C  A  I  A  E  F  Y  R  D  Q  G  E  H  A  L  I  I  Y  D  D  L  T  K  H  A  A  A  Y  R  Q

       *                *                   *               *                  *               960
TTATCTTTACTATTAAGACGCCCACTAGGACGTGAAGCCTTTCCAGGTGATGTCTTTTATGCTCATTCCAGACTTTTAGAAAGAGCATGCAAACTTAACACTAATTTCGGTGGAGGTTCT
L  S  L  L  L  R  R  P  L  G  R  E  A  F  P  G  D  V  F  Y  A  H  S  R  L  L  E  R  A  C  K  L  N  T  N  F  G  G  G  S

      *                  *                *                   *                   *           1080
TTAACCGCTTTACCTATTGTAGAAACACAAGCCGGTGATTTATCTGGTTATATCCCAACTAATATTATTTCTATTACTGATGGACAAATTTTCATGGAAAAAGATCTATTCTTTAAAGGC
L  T  A  L  P  I  V  E  T  Q  A  G  D  L  S  G  Y  I  P  T  N  I  I  S  I  T  D  G  Q  I  F  M  E  K  D  L  F  F  K  G

      *                   *                 *                  *                 *           1200
ATCCGTCCAGCTGTTAATGCTGGTAGTTCTGTTTCTAGAGTCGGTTCTAAAGCTCAACCATACGCTTTACGTTTAGTAACTGGTAATCTTCGTTATCAACTTGCTCAATATCGTGAATAT
I  R  P  A  V  N  A  G  S  S  V  S  R  V  G  S  K  A  Q  P  Y  A  L  R  L  V  T  G  N  L  R  Y  Q  L  A  Q  Y  R  E  Y

       *                  *                 *                  *                 *           1320
TCTGTCTTTGCACAATTTGATAATGATATCGATGATGTTACAAGAGGAATTCTTAATAGAGGTGCTTTATTAACTGAAATTCTTAAACAAGGTCCTAATATGCCAATGCAATTATATAAA
S  V  F  A  Q  F  D  N  D  I  D  D  V  T  R  G  I  L  N  R  G  A  L  L  T  E  I  L  K  Q  G  P  N  M  P  M  Q  L  Y  K

      *                   *                 *                    *                 *           1440
CAAGTTTTAATTATCTTAGCTGGTGCTTTTAATTTTATTACTCCTTTTATCAAAGATTTTAAAAAAGATTTGAGCAATATCTGTGCTTTATATGAAACTAAACTATTCAAGTTTACACAT
Q  V  L  I  I  L  A  G  A  F  N  F  I  T  P  F  I  K  D  F  K  K  D  L  S  N  I  C  A  L  Y  E  T  K  L  F  K  F  T  H

      *                  *                 *                    *                  *           1560
GATAAACCTGAACGAGCTGAATTATTCTATCCATTCTATGAATATTTATCATTCTTTGATAAAGCCGATTTTGGATTTGCTGACAATCCATTATTATTTATTTTAAGAAGTTTTGAACCG
D  K  P  E  R  A  E  L  F  Y  P  F  Y  E  Y  L  S  F  F  D  K  A  D  F  G  F  A  D  N  P  L  L  F  I  L  R  S  F  E  P

                 1590
GAATTCTATAAATCCATAAATGCCTAATAA
E  F  Y  K  S  I  N  A
```

Fig. 3 ── Sequence of the open reading frame coding for αATP synthase. The cDNA sequence of the alpha ATP synthase is shown from the ATG initiation codon (position 1) to the tandem termination codons (position 1590; Mahendran *et al.*, 1991). Below the DNA sequence is the amino acid sequence of the alpha subunit as deduced from the DNA sequence using the classic genetic code. Asterisks indicate cytidine residues which are not present in the analogous mtDNA sequence and create the open reading frame.

sequence similarity. Small regions of sequence identity with the cDNA were separated by -1 frameshifts. In the cDNA this frameshift was corrected in every case by the presence of a single cytidine residue. These cytidine residues disrupted some of the restriction sites present in the mtDNA and created others.

In order to complete the sequence of the mRNA for the alpha subunit, specific primers were synthesized which were complementary to sequences in the mtDNA. These sequences contained short stretches of open reading frame conserved in their inferred amino acid sequence. The primers were used to synthesize cDNAs which were subsequently amplified using the polymerase chain reaction. These cDNA amplification products were cloned in plasmids and used for sequence analysis. The sequence of these fragments completed a 528 codon reading frame with homology to the alpha subunit of ATP synthase from a number of organisms (Fig. 3). In all, 54 cytidine insertions were found in cDNAs relative to the mtDNA sequence. The insertions were fairly evenly distributed throughout the cDNA, the distance between the inserted cytidines ranging from 11 to 52 nucleotides with average of 26 nucleotides. Exceptions to this distribution pattern were located at cither end of the reading frame where stretches of 234 and 66 nucleotides lacked editing sites.

A major question was whether the frameshifted mtDNA sequence was the template for the mRNA or whether some related region with an intact reading frame coded for the mRNA. Although unprecedented, it was formally possible that the mtDNA sequence was a pseudogene and that a functional gene was present at some other location on the mtDNA or in the nucleus. Hybridization studies using the cDNAs as probes detected no other region of the mtDNA that was homologous to the cDNA. DNA from isolated nuclei was also devoid of homologous sequence. As an even more sensitive assay for sequences homologous to the cDNA, the polymerase chain reaction was used to search for sequences which lacked the frameshifts. To do this, pairs of short primers complementary to sequences with central cytidine insertions were synthesized. These primers failed to produce amplification products using as a template DNA isolated from purified nuclei, even though primers lacking the cytidine insertions were able to produce amplification products using as a template the trace amounts of contaminating mtDNA present in the nuclear DNA preparations. These results led us to conclude that the RNA was synthesized on the mtDNA template that we had cloned and scquenced and that cytidines were added by a co- or posttranscriptional RNA editing process to produce the open reading frame.

4.4 OTHER RNAs THAT REQUIRE EDITING

In order to determine whether RNAs other than the mRNA of the alpha subunit require editing, several other cDNAs were isolated and sequenced. These cDNAs were produced from rRNAs and tRNAs as well as mRNAs. Analysis of the mitochondrial small subunit rRNA indicated that it was also extensively edited (Mahendran *et al.*, 1992; Fig. 4). In the rRNA most of the editing sites consisted of single cytidine insertions. However, single uridines were inserted at three locations and four adenosines were inserted as a dinucleotide at two locations.

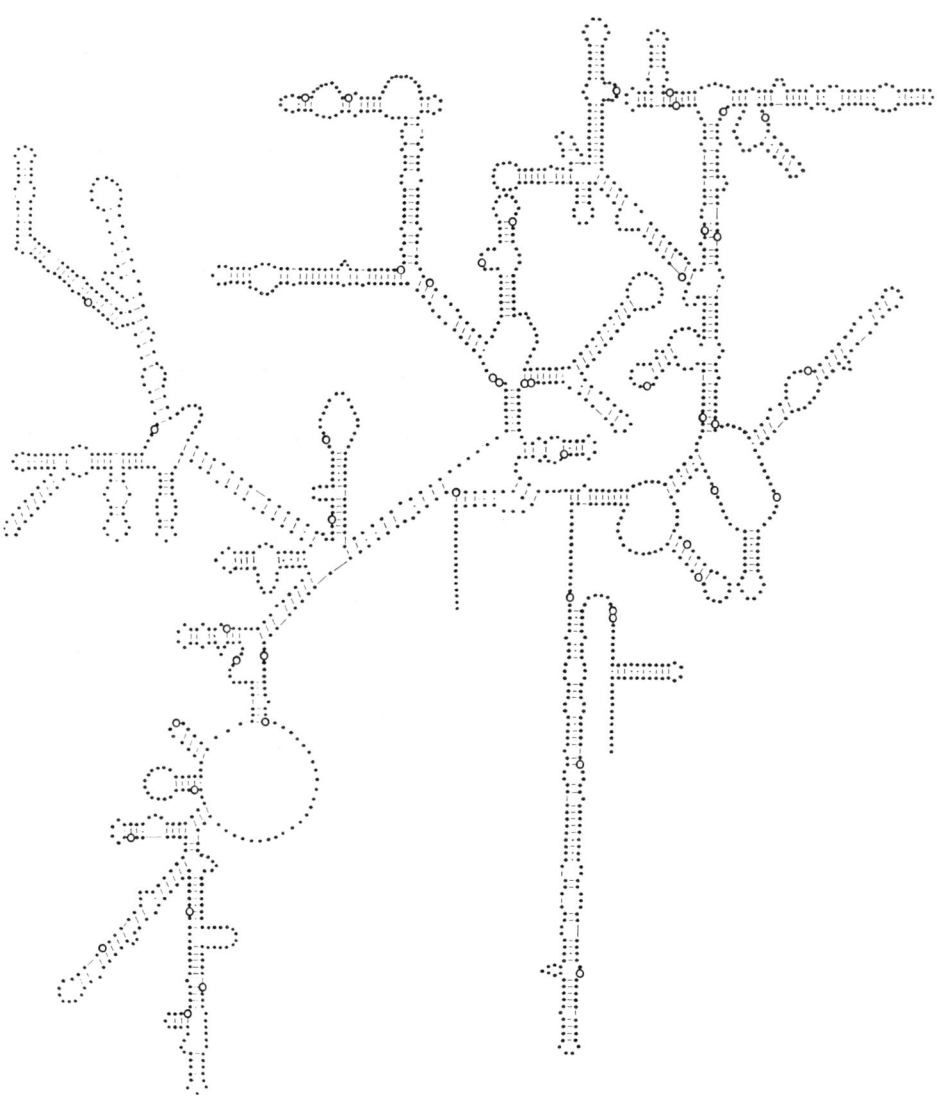

Fig. 4 — Small subunit rRNA from *Physarum* mitochondria. The secondary structure of the small subunit rRNA is presented as inferred from its primary sequence homology with the *E. coli* 16S rRNA. Closed circles indicate nucleotides encoded by the mtDNA; open circles indicate nucleotides added by editing. Lines between nucleotides indicate Watson-Crick base pairs. Dots between nucleotides indicate G:U base pairs. The secondary structure of the core region is modeled after the known secondary structure of the 16S rRNA of *E. coli*. Variable domains are depicted as one of several possible secondary structures.

The overall distribution of editing sites was similar to the editing in the alpha subunit of ATP synthase in that the insertions were distributed throughout the RNA separated by approximately 25 residues. However, this pattern was interrupted by several large gaps in which editing was absent, resulting in an overall average spacing of 40 nucleotides between insertions. Uridine insertions were intermixed with cytidine insertions and separated with the same general 25 nucleotide spacing. The one exception to this was an editing site in which a cytidine and uridine are inserted side-by-side. Since the secondary structure of the small subunit could be deduced by comparison with the 16S rRNA of *E. coli*, the location of editing sites could be analyzed in relation to the secondary structure. Editing sites do not correlate with any particular feature of the RNA and are not restricted to stems, loops, or single stranded regions. Also, editing sites are located both in highly conserved regions of the central core and in variable domains. Editing in conserved domains generally increases homology to conserved features, in some cases creating features conserved in nearly all small subunit rRNAs. Based on these results we expect that editing is required to produce functional rRNA.

A portion of the large subunit rRNA has been deduced from sequence analysis of its cDNA (Yang *et al.*, 1992). It is also extensively edited. As with the small subunit rRNA, the large subunit rRNA is predominantly edited with single cytidine insertions. However, two adenosine dinucleotide insertions and three additional mixed dinucleotide insertions have been identified. To our knowledge, the *P. polycephalum* rRNAs provide the first example of mixed nucleotide insertional editing.

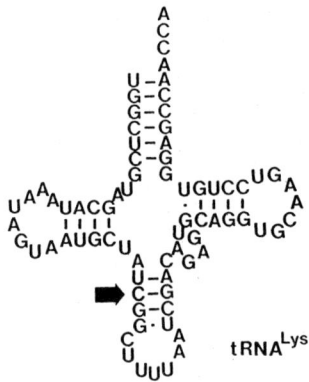

Fig. 5 — Cloverleaf secondary structure of tRNA^{Lys}. A 76 nucleotide RNA is shown in the classical cloverleaf structure of a tRNA. The CCA shown at the 3' end is assumed to be added posttranscriptionally. The tRNA is inferred to be tRNA^{Lys} based on the apparent UUU anticodon. The arrow indicates the location of a cytidine insertion which creates a G:C base pair in the anticodon stem.

Table 1 — Editing in mitochondrial RNAs of *Physarum*

RNA[a]	Inserted nucleotides		Insertion sites[b]		Spacing[c]
LSU rRNA (I)	C	50	C	48	46 ± 34
	U	4	AA	2	
	A	4	CU	2	
	G	1	GU	1	
			U	1	
SSU rRNA	C	41	C	40	40 ± 33
	U	3	AA	2	
	A	4	CU	1	
			U	2	
αATP Syn mRNA	C	54	C	54	26 ± 10
PL ATP Syn mRNA	C	9	C	9	24 ± 5
Cyt b mRNA (I)	C	34	C	31	27 ± 15
	U	8	CU	2	
	G	1	GC	1	
			U	6	
NADH SU1 mRNA	C	38	C	38	22 ± 8
	U	2	U	2	
tRNA[K]	C	1	C	1	
tRNA[E]	C	1	C	1	
	U	1	U	1	
tRNA[M]	C	1	C	1	

[a]LSU rRNA = large subunit rRNA, SSU rRNA = small subunit rRNA, αATP Syn = alpha subunit of ATP Synthase, PL ATP Syn = proteolipid subunit of ATP synthase, Cyt b = apocytochrome b, NADH SU1 = subunit 1 of the NADH dehydrogenase complex. (I) indicates that the sequence of the cDNA has not yet been completed.
[b]The number of sites with mono- or dinucleotide insertions relative to the mtDNA sequence. The identity of the insertion is indicated to the left.
[c]The average distance between insertion sites ± the standard deviation.

Having identified editing sites in both rRNA and mRNA we next examined cDNAs derived from several tRNAs. At least three tRNAs require editing to produce fully mature transcripts (Costandy *et al.*, 1992). In tRNA^{Lys} a cytidine insertion in the anticodon stem creates a G:C base pair (Fig. 5). In tRNA^{Glu} a uridine is inserted in the pseudouridine loop to create the GUUC sequence found in nearly all tRNAs and a single cytidine is inserted in the acceptor stem producing a G:C base pair. In tRNA^{Met} a cytidine is inserted into the anticodon stem creating an AluI site in its cDNA that is absent from the analogous region of the mtDNA.

Three additional mRNAs have also been analyzed. They encode the proteolipid subunit of ATP synthase (atp PL; Mahendran *et al.*, 1992), the apoprotein of cytochrome b (cyt b; Wang *et al.*, 1992), and subunit 1 of the NADH dehydrogenase complex (ND1; Ling and Miller, 1992). These mRNAs are predominantly edited with single cytidine insertions but the cytochrome b mRNA has six single uridine insertions. At three locations nucleotides are inserted side-by-side. These dinucleotides are GC and two CU. As of this writing we have identified 257 nucleotide insertions in nine RNAs; see Table 1. This indicates that RNA editing is a frequent event in *Physarum* mitochondria, and is a fundamental and necessary step in mitochondrial gene expression.

4.5 cDNAs WITH PARTIALLY EDITED SEQUENCES

A fundamental aspect of the mechanism of editing in *Physarum* is whether the editing occurs cotranscriptionally or posttranscriptionally. If editing is posttranscriptional, one may be able to find RNAs that are partially edited. These types of RNA have been identified in trypanosome mitochondria, although they have not been shown to be editing intermediates (Abraham *et al.*, 1988; Sturm and Simpson, 1990a; Stuart, 1990; see Chapters 2 and 3). The partially edited transcripts are characterized by being edited at the 3' end and unedited at the 5' end.

The abundance of unedited or partially edited sequences in *Physarum polycephalum* mitochondria is generally less than observed in trypanosome mitochondria. In *L. tarentolae* the abundance of edited transcripts in a steady-state population of RNAs varies with the RNA species. For example, 89% of the COII transcript is edited while only 36% of the MURF3 (ND7) transcripts are edited (Simpson and Shaw, 1989). In *Physarum*, sites in which editing creates or destroys restriction enzyme sites in the cDNAs have been exploited to assay editing frequency (Mahendran and Miller, unpublished data). PCR primers flanking but not overlapping known RNA editing sites were used to amplify cDNAs produced from mitochondrial RNAs. The amplification products produced were then digested to completion with a restriction enzyme whose site was created or destroyed by editing. Amplification products containing restriction sites which are created by editing were about 95% digested; amplification products containing restriction sites destroyed by editing were about 5% digested. These results suggest that the majority of the steady state RNA in *Physarum* is edited. The amount of unedited RNA could even be lower than 5%,

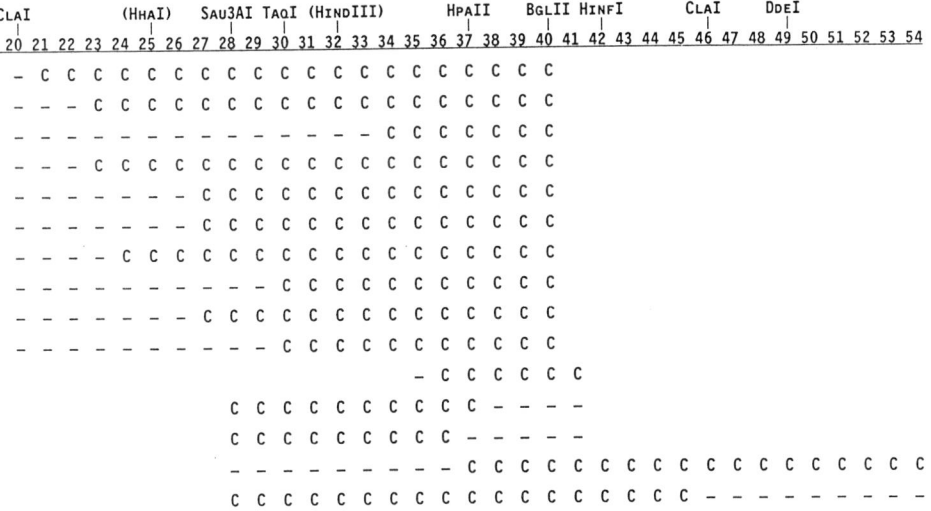

Fig. 6 — cDNAs with partially edited sequences. The numbered upper line indicates editing sites 20 to 54 of the mRNA for the alpha subunit of ATP synthase. Restriction enzyme sites destroyed (parentheses) or created (no parentheses) by cytidine insertion are indicated above the line. Fifteen cDNAs with a partially edited sequence are represented below the line. Insertion sites with a cytidine insertion are designated with a C; potential insertion sites which lack a cytidine are indicated with a -. The upper ten cDNAs were produced by using primers which would select for 3' edited, 5' unedited cDNAs (see text). The lower five cDNAs were obtained without selection.

if traces of DNase resistant mtDNA (with the <u>un</u>edited sequence) contaminated the RNA preparations.

Because partially edited transcripts are present in low amounts in *Physarum* mitochondria, we chose to use PCR amplification to determine whether they resemble the partially edited transcripts in trypanosome mitochondria. Amplification of cDNAs using primers which did not overlap with editing sites occasionally produced amplification products which were partially edited. Similar to the results obtained for trypanosome mitochondrial RNA, edited sites were grouped together and separated from unedited sites by a single boundary. In contrast to the editing pattern in trypanosome mitochondria, however, the edited region of partially edited cDNAs could be located at either the end corresponding to the 5' or the 3' end of the RNA (Fig. 6). Also, no unusual boundary sequences have been observed in these cDNAs. The frequency of cDNAs with partially edited sequences can be increased by using primers that overlap editing sites. Primer sets with one primer complementary to edited sequences and one primer complementary to unedited sequences select for partially edited cDNAs. The partially edited cDNAs produced in this way have the same characteristics as partially edited cDNAs produced without selection. These observations could mean that RNA editing in *Physarum* mitochondria is posttranscriptional and processive but bidirectional. Although at this stage we cannot completely rule out experimental artifacts, it is clear that similar experimental protocols do not reproduce the unidirectional nature of editing observed in trypanosomes.

4.6 FEATURES OF EDITING SITES IN *PHYSARUM* MITOCHONDRIA

Several features make *Physarum* editing different from the insertional editing found in trypanosomes. First, and most obvious, is the fact that the majority of inserted nucleotides in *Physarum* are cytidines while editing in the trypanosomes exclusively involves uridines. While 90% of the editing sites are cytidine insertions, a significant portion are uridine insertions (7%), the remainder being guanosine and adenosine insertions (Table 1). Editing sites in trypanosomes consist of insertions of 1 to 8 uridines per site or the occasional deletion of 1 to 5 uridines (Benne, 1989). In *Physarum* the majority of insertions (96%) are single nucleotides and the remainder are dinucleotides. No deletions or substitutions have been observed.

The distribution of editing sites within an RNA in *Physarum* mitochondria is also different from that in trypanosomes. In trypanosomes editing sites are often clustered in a single region of the RNA, e.g. in the transcripts for cox2 editing sites are confined to a 4 nucleotide region internal to the reading frame and in the transcripts for cyt b and MURF 2, editing sites are confined to a 20 to 25 nucleotide region at the 5' end. In contrast, some RNAs (e.g. those for cox3, ND7 and atp6) are edited so extensively that the reading frame is essentially created *de novo* (Benne, 1989; Simpson and Shaw, 1989; Stuart, 1990; see Chapters 2 and 3).

Neither of these editing patterns is seen in *Physarum* mitochondria. The distribution of editing sites in the mitochondrial RNA of *Physarum* is noticeably uniform. As mentioned above, editing is generally distributed throughout the entire RNA and the editing sites are spaced at an average of about 25 nucleotides in mRNA and 43 nucleotides in rRNA (Table 1). So far, no two editing sites have been observed closer than nine nucleotides. The variation around the average spacing appears to be random. The exception to this is dinucleotide insertions where two residues are inserted side-by-side. The sites of dinucleotide insertions generally conform to the spacing of single nucleotide insertions. Although cytidines and uridines are inserted as single nucleotides, they are never inserted as CC or UU dinucleotides but are invariably mixed in composition (CU, GU, or GC). Conversely, adenosines are inserted as AA dinucleotides but not as single As.

In *Physarum*, editing of mRNAs creates codons to produce a single continuous reading frame. Neither the codons created by editing nor the location of the insertion within the codon is random. Table 2 shows the 140 codons created by cytidine or uridine insertion. Of the possible 37 codons that can be created by cytidine insertion only 27 have so far been observed. The most

Table 2 — Codon use and codons created by editing[a]

TTT Phe	94	TCT Ser	38 (1)	TAT Tyr	58	TGT Cys	9
TTC Phe	20 (5)	TCC Ser	8 (5)	TAC Tyr	4 (2)	TGC Cys	4 (1)
TTA Leu	103(2)	TCA Ser	10 (2)	TAA End	0	TGA End	0
TTG Leu	11	TCG Ser	2 (1)	TAG End	0	TGG Trp	16
CTT Leu	40 (7)	CCT Pro	32 (5)	CAT His	15	CGT Arg	23 (1)
CTC Leu	5 (2)	CCC Pro	4 (3)	CAC His	0	CGC Arg	6
CTA Leu	16 (6)	CCA Pro	19 (6)	CAA Gln	39(5)	CGA Arg	7 (1)
CTG Leu	1 (1)	CCG Pro	5 (2)	CAG Gln	2	CGG Arg	0
ATT Ile	71	ACT Thr	34 (3)	AAT Asn	47	AGT Ser	16
ATC Ile	31(27)	ACC Thr	15(13)	AAC Asn	4 (2)	AGC Ser	7 (2)
ATA Ile	24	ACA Thr	17	AAA Lys	40	AGA Arg	12
ATG Met	22	ACG Thr	0	AAG Lys	4	AGG Arg	0
GTT Val	41	GCT Ala	60 (3)	GAT Asp	47	GGT Gly	62
GTC Val	17(16)	GCC Ala	24(14)	GAC Asp	4 (1)	GGC Gly	4
GTA Val	23 (1)	GCA Ala	17	GAA Glu	47	GGA Gly	22
GTG Val	2	GCG Ala	2	GAG Glu	4	GGG Gly	3

[a]The cognate amino acids for the codons were assigned using the classic genetic code. The number to the right of the amino acid is the total codon use in the four mRNAs listed in Table 1. The numbers in parentheses are the number of codons created by cytidine or uridine insertions.

Table 3 — Dinucleotides upstream of cytidine insertions[a]

AA	7	AG	14	AT	58
GA	5	GG	3	GT	26
CA	1	CG	0	CT	8
TA	7	TG	4	TT	9

[a]Composite frequency of the dinucleotide immediately upstream of cytidine insertions in the RNAs listed in Table 1. Dinucleotides with C in the second position have been eliminated owing to the inherent ambiguity of the insertion position.

common codons created are ATC, GTC, ACC and GCC, which constitute 50% of the codons created by cytidine insertion. Of the 85 codons in which the location of the cytidine insertion can be unambiguously assigned, 67% are inserted in the third (wobble) position, significantly larger than the 33% expected by random insertion.

We have carefully analyzed the sequence context of the editing sites to determine if a feature of the primary sequence is involved in determining the editing site location. Although here again a bias in the location of editing sites is observed, it is not sufficient to specify editing sites. As an example, Table 3 shows the distribution of dinucleotides located immediately upstream of editing sites. Purine-pyrimidine dinucleotides are located upstream of the cytidine insertion in 60% of the sites. Since the exact location of the cytidine insertion is ambiguous at locations where a cytidine dinucleotide results, these sites were not considered in the analysis. While there is a marked preference for purine-pyrimidine dinucleotides upstream of cytidine, many editing sites are located downstream of other dinucleotides and many purine-pyrimidine dinucleotides are not followed by editing sites. Clearly, this bias is not sufficient to explain editing site specificity.

4.7 POSSIBLE MECHANISMS OF RNA EDITING IN *PHYSARUM*

With over 250 editing sites identified, *Physarum* mitochondria display the most complex system of insertional editing outside of the Trypanosomatidae. The lack of any apparent consensus sequence or secondary structure to define the location of editing sites leads one to conclude that the information defining editing site location is not resident in the edited RNA itself. The discovery in trypanosomes of antisense RNAs complementary to edited sequences (guide RNAs) has led to the proposal that these guide (g) RNAs are the informational molecules which specify editing sites (Blum *et al.*, 1990). These gRNAs are about 60 nucleotides in length (Blum *et al.*, 1990; Blum and Simpson, 1990) and are encoded both by the mitochondrial maxicircle DNA (Blum *et al.*, 1990; Van der Spek *et al.*, 1991) and by minicircle DNA (Sturm and Simpson, 1990b; Pollard *et al.*, 1990; Koslowsky *et al.*, 1990; Bhat *et al.*, 1990). Although different models of how editing sites are specified in trypanosomes have been proposed (Blum *et al.*, 1990; Decker and Sollner-Webb, 1990), all have in common the premise that the editing site is identified through a mismatch between the gRNA and the unedited mRNA. This mismatch is ultimately removed by the insertion or deletion of one or more uridines. In this respect the gRNA acts as a guide to identify the editing site but not as a template to specify the nucleotide inserted or deleted; for more information, see Chapters 2 and 3.

It is tempting to propose that nucleic acids analogous to gRNAs are central to the mechanism of editing in *Physarum* mitochondria given the complexity of the editing process. However, while guide RNA models would suffice to explain editing site selection, they do not explain the nucleotide specificity in a system with mixed nucleotide insertional editing. Either a template must be present that can specify nucleotides (by Watson-Crick base pairing, for example), or different types of guide molecules must exist which are specific both for a site and for the nucleotide to be inserted. In either case the most likely molecule to contain the information for editing is a nucleic acid antisense to the edited RNA. To this end we have screened *Physarum* mitochondrial RNA with probes which would hybridize with antisense RNA. These probes have reproducibly failed to detect antisense RNA in mitochondrial RNA preparations. This failure may be the result of extensive G:U base pairing between the antisense RNA and edited RNA which results in the sequence of the antisense RNA being significantly different from that predicted by Watson-Crick base pairing. In typanosomes, numerous G:U base pairs are predicted between gRNAs and their mRNA (Blum *et al.*, 1990; Sturm and Simpson, 1990b; Pollard *et al.*, 1990; Koslowsky *et al.*, 1990; Bhat *et al.*, 1990; Van der Spek *et al.*, 1991). Degenerate probes designed to detect antisense RNAs which could hybridize with the edited RNA through either G:U or Watson-Crick base pairing have also failed to detect any RNA. Computer searches designed to detect regions of the mtDNA which might code for the antisense RNAs have not been fruitful either, but only about 40% of the mtDNA has been sequenced, so portions of the mtDNA could still contain clusters of genes coding for antisense RNAs that can not be detected by hybridization.

Guide RNAs have been shown to possess oligouridine tails at their 3' end (Blum and Simpson, 1990). Cech (1991) and Blum *et al.* (1991) have proposed that these oligouridine tails participate in the mechanism of editing via a two-step transesterification reaction in which uridines in the tail are transferred to the editing site. By analogy one might expect oligocytidine residues to be found at the 3' end of RNAs involved in editing in *Physarum* mitochondria. However, no mitochondrial RNAs with 3' oligocytidine sequences have been detected.

4.8 PERSPECTIVE

Sequence analysis of the mtDNA and analogous cDNAs has led to the identification of over 250 editing sites in nine genes clustered within one quadrant of the mtDNA. The transcription pattern of the mtDNA (Jones *et al.*, 1990) leads us to predict that at least an equal number of genes are yet to be identified. Unless the density of editing in the RNAs produced from this quadrant is atypical of the entire genome, the insertion of nucleotides at over 1000 different sites may be required for mitochondrial biogenesis in *Physarum*. The relatively uniform distribution of editing sites over the length of the RNA requires that if antisense nucleic acids produce editing site specificity, as is likely in the trypanosomes, their complexity must be nearly equal to the RNAs that are edited. Since *Physarum* mitochondria lack DNA analogous to minicircles, it is likely that information for editing specificity is ultimately resident in the mtDNA. Further analysis of the mtDNA and the RNAs it encodes should provide insights into the mechanism of editing.

The failure to detect RNAs with gRNA properties and the numerous differences in the features of editing between *Physarum* and trypanosomes lead to the prospect that these two processes have fundamentally different mechanisms. Further understanding of the mechanism of editing in *Physarum* mitochondria will probably require the development of rapid and reliable biochemical assays for editing and the development of an *in vitro* editing system.

REFERENCES

Abraham, J.M., Feagin, J.E. & Stuart, K. (1988). Characterization of cytochrome c oxidase III transcripts that are edited only in the 3' region. *Cell* **55** 267-272.

Benne, R. (1989) Biochim. Biophys. Acta **1007**: 131-139.

Bhat, G.J., Koslowsky, D.J., Feagin, J.E., Smiley, B.L. & Stuart, K. (1990). An extensively edited mitochondrial transcript in kinetoplastids encodes a protein homologous to ATPase subunit 6. *Cell* **61** 885-894.

Blum, B., Bakalara, N. & Simpson, L. (1990). A model for RNA editing in kinetoplastid mitochondria: "guide" RNA molecules transcribed from maxicircle DNA provide the edited information. *Cell* **60** 189-198.

Blum, B. & Simpson, L. (1990). Guide RNAs in kinetoplastid mitochondria have a nonencoded 3' oligo-(U) tail involved in recognition of the pre-edited region. *Cell* **62** 391-397.

Blum, B., Sturm, N.R., Simpson, A.M. & Simpson, L. (1991). Chimeric gRNA-mRNA molecules with oligo(U) tails covalently linked at sites of RNA editing suggest that U addition occurs by transesterification. *Cell* **65** 543-550.

Cech, T.R. (1991). RNA editing: world's smallest introns. *Cell* **64** 667-669.

Costandy, H., Mahendran, R., Spottswood, M.R., McCoy, H. & Miller, D.L. (1993) Manuscript in preparation.

Daniel, J.W. & Baldwin, H.H. (1964) Methods of culture for plasmodial myxomycetes. *Methods Cell Physiol.* **1** 9-41.

Decker, C.J. & Sollner-Webb, B. (1990). RNA editing involves indiscriminate U changes throughout precisely defined editing domains. Cell 61, 1001-1011.

Evans, T.E. & Suskind, D. (1971) Characterization of the mitochondrial DNA of the slime mold *Physarum polycephalum. Biochim. Biophys. Acta* **228** 350-364.

Hasegawa, M., Iida, Y., Yano, T., Takaiwa, F. & Iwabuchi, M. (1985) Phylogenetic relationship among eukaryotic kingdoms inferred from ribosomal RNA sequences. *J. Mol. Evol.* **22** 32-38.

Johansen, T., Johansen, S. & Haugli, F. (1988) Nucleotide sequence of the *Physarum polycephalum* small subunit ribosomal RNA as inferred from the DNA sequence: secondary structure and evolutionary implications. *Curr. Genet.* **14** 265-273.

Jones, E.P., Mahendran, R., Spottswood, M.R., Yang, Y.-C. & Miller, D.L. (1990) Mitochondrial DNA of *Physarum polycepahlum*: physical mapping, cloning and transcription mapping. *Curr. Genet.* **17** 331-337.

Koslowsky, D.J., Bhat, G.J., Perrollaz, A.L., Feagin, J.E. & Stuart, K. (1990). The MURF3 gene of *Trypanosoma brucei* contains multiple domains of extensive editing and is homologous to a subunit of NADH dehydrogenase. *Cell* **62** 901-911.

Ling, M.-L. and Miller, D.L. (1993) Manuscript in preparation.

Mahendran, R. (1990) Structure and expression of protein-coding genes on the mitochondrial DNA of *Physarum polycephalum.* Ph.D. Thesis, The University of Texas at Dallas.

Mahendran, R., Spottswood, M.R. & Miller, D.L. (1991). RNA editing by cytidine insertion in mitochondria of *Physarum polycephalum. Nature* **349** 434-438.

Mahendran, R., Spottswood, M.R., Ghate, A., Jeng, K. & Miller D.L. (1993) Editing of the mitochondrial small subunit rRNA in *Physarum polycephalum.* Manuscript submitted.

Mahendran, R., Rankins, R. & Miller, D.L. (1993) Manuscript in preparation.

Margulis, L. & Schwartz, K.V. (1988) Five Kingdoms: an illustrated guide to the phyla of life on earth. Freeman Press, New York.

Pollard, V.W., Rohrer, S.P., Michelotti, E.F., Hancock, K. & Hajduk, S.L. (1990). Organization of minicircle genes for guide RNAs in *Trypanosoma brucei. Cell* **63** 783-790.

Simpson, L. & Shaw, J. (1989). RNA editing and the mitochondrial cryptogenes of kinetoplastid protozoa. *Cell* **57** 355-366.

Stuart, K. (1991). RNA editing in trypanosomatid mitochondria. *Annu. Rev. Microbiol.* **45** 327-344.

Sturm, N.R. and Simpson, L. (1990a). Partially edited mRNAs for cytochrome b and subunit III of cytochrome oxidase from *Leishmania tarentolae* mitochondria: RNA editing intermediates. *Cell* **61** 871-878.

Sturm, N.R. & Simpson, L. (1990b). Kinetoplast DNA minicircles encode guide RNAs for editing of cytochrome oxidase subunit III mRNA. *Cell* **61** 879-884.

Van der Spek, H., Arts, G.-J., Zwaal, R.R., Van den Burg, J., Sloof, P. & Benne, R. (1991). Conserved genes encode guide RNAs in mitochondria of *Crithidia fasciculata.* EMBO J. 10, 1217-1224.

Wang, S., Mahendran, R., Volpert, A. & Miller, D.L. (1993) Manuscript in preparation

Yang, N., Costandy, H., Spottswood, M.R. & Miller, D.L. (1993) Manuscript in preparation.

5

PARAMYXOVIRUS P GENE mRNA EDITING

Daniel Kolakofsky, Joseph Curran, Thiery Pelet and Jean-Philippe Jacques
Department of Genetics and Microbiology, University of Geneva School of
Medicine CMU, 9 Avenue de Champel, CH-1211 Geneva, Switzerland

5.1 A BRIEF OVERVIEW OF PARAMYXOVIRUS RNA SYNTHESIS

Paramyxoviruses contain non-segmented ssRNA genomes of around 15 kb and
negative polarity, i.e. complementary to the viral mRNAs. These genomes (and
negative-strand viral genomes in general) function not as free nucleic acid, but as
helical nucleocapsids (NC) assembled with the viral nucleoprotein (NP) in which
the genome RNA represents only 4% by weight. It is these nucleocapsids which
are the templates for RNA synthesis. The viral polymerase is composed of two
subunits, the phosphoprotein P and the large protein L, but its overall structure is
unclear.

The genome contains about 6 genes or transcriptional units, preceded by a
short non-coding leader region. Between the genes are highly conserved
sequences thought to function as the mRNA polyadenylation/termination and the
mRNA start signals, separated by a short non-transcribed spacer. A central
feature of (-)RNA genomes is that they are templates for the synthesis of both
mRNA (transcription) and the full-length antigenomes (the complementary
intermediates in genome replication). During transcription, the polymerase starts
at the 3' end of the template and sequentially produces the leader RNA and
mRNAs by stopping and restarting at each of the junctions.

The mRNAs contain a poly(A) tail of several hundred nucleotides and the
start of this tail maps to the U_{5-7} stretch within the conserved termination
sequence of the (-) template (e.g., 3'AUUC<u>UUUUU</u> for Sendai virus). The
poly(A) tail is presumably made by polymerase slippage and reiterative copying
of the U_{5-7} stretch (McGeoch, 1979; Iverson and Rose, 1981; Schubert *et al.*,

1980; Robertson *et al.*, 1981; Giorgi *et al.*, 1983; Gupta and Kingsbury, 1984), and this phenomenon can be referred to as "stuttering". The same polymerase is thought to start the next mRNA only after terminating poly(A) synthesis. During antigenome synthesis or replication, the polymerase reads through all junctions to make an exact full-length complementary copy. Unlike the mRNAs, but like genomes, the antigenomes are assembled into NCs and this takes place concomitantly with their synthesis. (Antigenomes presumably also function as templates only in the assembled form.) This coupling of antigenome synthesis and assembly is thought to play a key role in deciding whether the junctional stop/restart signals are obeyed. For a more complete review, see Banerjee (1987), and Kolakofsky and Roux (1987).

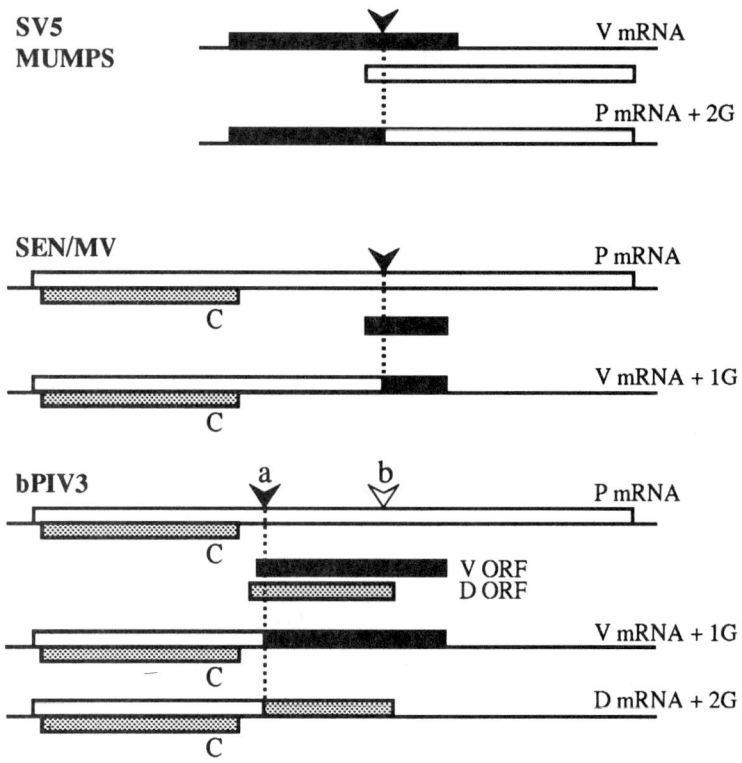

Fig. 1 —— Schematic representation of the paramyxovirus P gene mRNAs. The mRNAs are indicated as lines and the ORFs as boxes. For each group, the upper line shows the mRNA which is an exact copy of the gene, and the beginning of the ORF box indicates the ribosomal start codon. When more than one ORF box is attached to the line, they are accessed by alternate initiation codons. The boxes below indicate alternate downstream ORFs which are accessed by G insertions in the mRNAs. The positions of the insertions are shown by the arrowheads and the dotted vertical lines. The position of 2 putative insertion sites of hPIV3 has been indicated (a and b). bPIV3 may only use site a (for details see Section 5.7). The three possible ORFs are indicated by open, stippled, and black boxes.

Except for their poly(A) tails, most mRNAs are faithful copies of their genes. This is not so, however, for the P gene mRNA(s), the subject of this review.

5.2 THE DISCOVERY OF G INSERTIONS WITHIN THE P GENE mRNAs

Even before the realization that a form of mRNA editing operated in paramyxoviruses, their P genes were considered unusual. In every case multiple proteins were expressed from this gene, in contrast to all the other viral genes in which only a single primary translation product was expressed. The Paramyxoviridae is a relatively large family of viruses, and their P genes can be divided into basically two groups (see Fig. 1): i) the Sendai/measles virus or SEN/MV group, which express longer P proteins (568-603 amino acids) and which also contain an alternate ORF to express the C protein(s); the C protein ORF overlaps the amino-terminal sequences of the P protein ORF in the +1 frame, and ii) the simian virus 5 (SV5)/mumps virus group, which express shorter P proteins (245-400 amino acids) and do not contain the overlapping C protein ORF.

For the SEN/MV group, there was evidence that the P and C proteins were translated from the same mRNA by ribosomal choice, via leaky ribosomal scanning and sometimes by what appears to be a scanning-independent mechanism (Curran and Kolakofsky, 1988, 1989). However, even for the SV5/mumps group, which did not contain the alternate C protein ORF, these P genes also expressed in at least two proteins. For example, the mumps virus P gene also expresses the non-structural proteins NS1 and NS2 (since renamed V and I, with tongue in cheek, for consistency with SV5) (Herrler and Compans, 1982; Paterson and Lamb, 1990). However, until the sequence of the first P gene in this group was determined (correctly), there was no way of knowing what mechanisms might be operating to produce these multiple proteins.

The first such gene determined was that of SV5 (Thomas et al., 1988) (Fig. 1). The sequence was both determined from three separate cDNA clones made from mRNA and then confirmed on the viral genome. This yielded the surprising result that there was no single ORF large enough to code for the SV5 P protein. Rather, the gene contained two separate ORFs at each end, which briefly overlapped in the middle. mRNA transcribed from these cDNAs *in vitro* could be translated into the shorter V protein encoded by the 5' proximal ORF, but not into the longer P protein. After a period of frustration, the solution to this dilemma became clear when Thomas et al. prepared a second cDNA library with a larger number of P gene clones, and sequenced 22 of these. Twelve of these were identical to the genome and the 3 mRNA clones sequenced initially. The ten other sequences were identical to these, except that they had 2 G residues inserted within a short run of Gs in a region where the 2 ORFs overlapped (arrowhead in Fig. 1). This fused the separate ORFs at each end into a single continuous unit, and mRNAs prepared from these latter clones translated into P (but not V) proteins *in vitro*.

A second unexpected finding emerged from this study. Whereas the P protein sequences are the least conserved of the viral proteins between different paramyxoviruses, that region of V which is not shared with P (i.e., the

C-terminal region) is cysteine-rich and among the best conserved of all the viral sequences. This cys-rich V domain was in fact present in all the P genes of the SEN/MV group previously sequenced, but had not been noticed because it was relatively short (ca. 75 aa), in the middle of the genes, and generally lacked a start codon. Thomas *et al.* (1988) predicted that these other viruses would also express V proteins by a similar mechanism, except that in the SEN/MV group 1 G would need to be inserted to switch from the P to the V ORF.

This prediction was verified, in fact, before the paper of Thomas *et al.* appeared in print. M. Billeter's laboratory in Zurich, who were studying the mutations generated in the various MV genes during persistent infections of human brain (known as subacute sclerosing panencephalitis or SSPE), independently found that some P gene mRNAs generated from cDNA clones made normal-sized P proteins, whereas others made proteins of only half this size. When the various cDNAs were sequenced, the only real difference between them was that those that made the shorter V protein contained a single extra G within a short run of Gs, which was also absent in the genome (Cattaneo *et al.*, 1989). Once it became clear what to look for, other examples of G insertions in P gene mRNAs then followed shortly (Sendai virus, Vidal *et al.*, 1990a; mumps virus, Paterson and Lamb, 1990; bPIV3, Pelet *et al.*, 1991; hPIV3, Galinski *et al.*, 1992; LPMV, Berg *et al.*, 1992).

5.3 THE MECHANISM OF THE G INSERTIONS

As mentioned above, poly(A) tail formation of these mRNAs is thought to take place by the viral polymerase reiteratively copying the U_{5-7} stretch within the termination sequence at the end of each gene. Since the G insertions appear to take place only in the P gene mRNAs and not in the P genes (i.e., during transcription but not during genome replication), and because there was also some homology between the sequences which precede the U_{5-7} stretch and those of the G insertion site, it seemed reasonable that the G insertions could arise by reiterative copying of the short C stretch on the genome template at the insertion site (Thomas *et al.*, 1988; Cattaneo *et al.*, 1989).

Support for this notion was obtained with Sendai (SEN) virus (Vidal *et al.*, 1990a). All paramyxovirions contain a template bound-polymerase and purified virions can make mRNA *in vitro*. However, only for SEN had the conditions for efficient mRNA synthesis been worked out. When P gene mRNAs made with purified SEN virions were examined and compared to those made *in vivo*, they were found to be "edited" very similarly; approximately 25% had a single G insertion, and about 2% had insertions which ranged from 2 to 8 Gs. This suggested that the insertions were due to viral proteins. Equally of importance, when the P gene mRNAs were expressed from recombinant vaccinia viruses in cells doubly infected with SEN, those made from the SEN genome were modified as before, whereas those made from the vaccinia genome were not modified at all. The inability of the insertion activity to act *in trans* supports the idea that the insertions occur during mRNA synthesis, rather than on preformed mRNA. The G insertions also do not occur when the mRNA is made *in vivo* from plasmid DNA via T7 polymerase (Curran *et al.*, 1991), or from an RNA vector which uses the Sindbis virus polymerase (U. Gegenmuller and S. Schlesinger, unpublished). Sindbis virus, however, is a positive strand RNA virus, uses "naked" RNA as template, and does not form poly(A) tails by a stuttering mechanism. All these data, of course, are consistent with a co-transcriptional mechanism.

On the other hand, it is entirely possible that when a portion of the SEN P gene containing the editing site is placed in an artificial vector system based on another negative strand RNA virus, such as influenza virus or a rhabdovirus, G insertion at this site in the mRNAs will take place. Although influenza virus and rhabdoviruses are not known to edit their own mRNAs, they nevertheless use a nucleocapsid as template for RNA synthesis, and also form poly(A) tails on their mRNAs by stuttering on an oligo(U) stretch.

5.4 A STUTTERING MODEL FOR THE G INSERTIONS

How, then, would this stuttering or reiterative copying proposed for the G insertions take place? The mechanism implies that the viral polymerase reiteratively copies one of the three C residues on the Sendai genome template (nt 1051-1053). This presumably occurs when the 3' end of the nascent mRNA at the insertion site, which is base-paired to the template over a few nucleotides, slips backwards or upstream on the template together with the polymerase before the next base is incorporated (right side, Fig. 2). The frequency with which Gs are inserted into the P mRNAs appears to be tightly controlled. For Sendai virus *in vivo*, 31 ± 2% of the mRNAs contain a 1 G insertion (5 determinations, Vidal *et al.*, 1990a). It seems reasonable that the polymerase would pause at the insertion site, otherwise there would not be time for the slippage to occur. RNA polymerases are known to pause during transcription on both procaryotic and eucaryotic DNA not only at termination sites, but well within the mRNA (Von Hippel *et al.*, 1984; Platt, 1986; Reines *et al.*, 1987), and pause times of 10 s to several minutes have been estimated (Chamberlin, 1976; Krakow *et al.*, 1976). The reasons for the polymerase pausing are unclear, but pausing occurs *in vitro* even in the presence of high NTP concentrations (Von Hippel *et al.*, 1984), and for RNA polymerase II the pausing is independent of the structure of the nascent mRNA (Reines *et al.*, 1987). These findings suggest that pausing is an intrinsic property of the template at these sites.

Assuming that the length of the pause is described by a bell-shaped curve (Fig. 3), a plausible mechanism to explain how the frequency of inserted mRNAs is controlled can then be based on this distribution. If, for example, the average pause time was 10 s and the minimum time necessary for slippage were slightly longer (e.g. 13 s), then insertions would occur at a fixed frequency of less than one. The fraction of mRNAs with G insertions is then determined by the fraction of the polymerases which pause longer than the time required for slippage, and changes in either of these two parameters would alter the frequency of G insertions. Conditions which increase the pause, for example, by so limiting the NTP required to continue elongation that further incorporation becomes rate limiting (Ruterhouser and Richardson, 1989), might then increase the frequency of G insertions.

We therefore examined the effect of limiting NTP concentrations during Sendai virus mRNA synthesis on the frequency of G insertions, using virion polymerase reactions. CTP, UTP, or GTP were individually lowered from 1 mM to 25 µM (and mRNA synthesis decreased about 9-fold), and the frequency of G insertions were determined by a cloning/oligonucleotide-typing method (Vidal *et al.*, 1990a) (Table 1). ATP was not lowered, since all viral RNAs start with ATP and its relatively high K_m for synthesis (300 µM) reflects initiation rather than internal incorporation. Neither the frequency of single G insertions nor that of

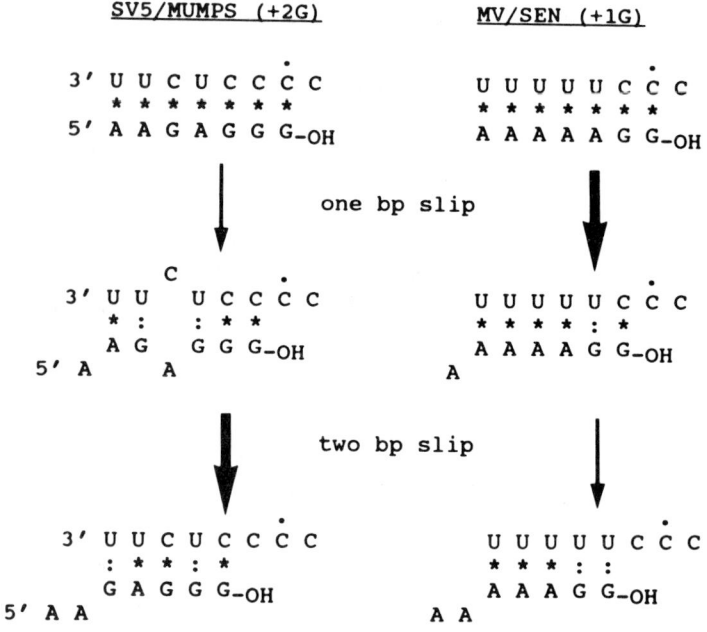

Fig. 2 — Polymerase and nascent chain slippage at the editing sites. The proposed events at the insertion site for each virus group are shown. The top line in each case is the genome, and only the consensus sequence is shown, with a dot above marking the proposed pause site. The bottom line is the nascent mRNA, with its 3'-OH end indicated. Only 7 bp between the mRNA and the template are shown, with normal pairs indicated with asterisks, and U:G pairs with 2 dots. The top duplexes show the nascent chains at the pause site before slippage, the middle and bottom duplexes after a one and two bp slippage, respectively. The thickness of the arrows indicates the preference of a one or two base pair initial slippage for each group, based on the stabilities of the misalignment intermediates. Incorporation after slippage which brings the polymerase back to the pause site leads to the G insertions, and incorporation beyond the pause site fixes the events.

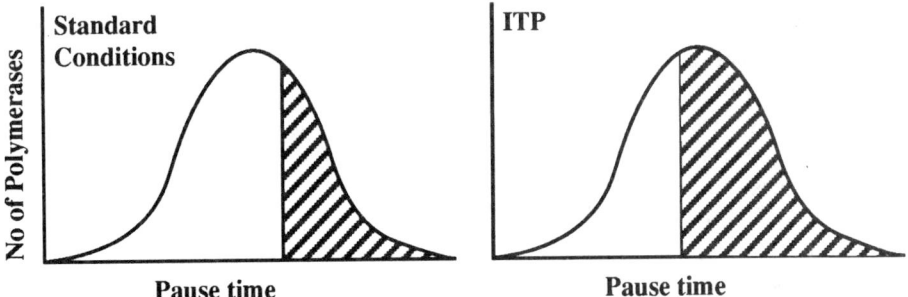

Fig. 3 — Polymerase pausing and the frequency of G insertions. The distribution of the length of the pause at the insertion site (before slippage/insertion) for individual polymerases is shown as a bell shaped curve. The vertical line within the curve indicates the minimum time necessary for slippage to occur, and the shaded area to the right indicates the fraction of polymerases which will insert one or more Gs. The difference between the fraction of inserted mRNAs *in vivo* (31% ± 2%) and under standard conditions *in vitro* (20% ± 2%) could be due to a difference in either of these two parameters. The effect of replacing guanosine with inosine (right side) is shown as lowering the minimum time required for slippage.

multiple G insertions was basically different from the standard reaction when either CTP or UTP was decreased to 25 µM (Table 1). When treated as a group, 19.9% ± 1.7% contained +1 G, and 2.6% ± 0.5% contained >1 G. Only when GTP was lowered to 25 µM were significant differences apparent, but in an unexpected fashion. The frequency of single G insertions was decreased by half, whereas those with multiple G insertions were increased 3-fold. The net result, however, was that the total fraction of inserted mRNAs varied little, if at all.

In terms of the stuttering model, we did not expect any differences at 25 µM UTP, as the first U after the G run is 3 nt downstream on the mRNA (5'GGGCA<u>U</u> 3'). Causing the polymerase to wait longer before it can incorporate this base should have no effect because it is too far from the insertion site. By the same criteria, the lack of any effect at 25 µM CTP (5'GGG<u>C</u>AU 3') would argue that if the stuttering mechanism applies, stuttering would not take place at the third C on the template, but presumably on the first or second C (3'<u>CC</u>CGUA 5'). If this were so, at low GTP there might be more time for slippage before the second or third G was incorporated (5'G<u>GG</u>CAU 3'), and so the frequency of insertions should increase. This is in essence what has occurred, even though the total fraction of inserted mRNAs is unchanged. Under standard conditions one G is added 18.5% of the time, and an average of six Gs are added at 2.7%, such that overall an average of 1.6 Gs are added when insertions occur. The same calculation when 25 µM GTP is used, shows that an average of 3.6 Gs are added when insertions occur. In terms of the stuttering model, it would appear that 25 µM GTP is not low enough to lengthen significantly the initial pause at the insertion site, and so the total fraction of inserted mRNA is unchanged. However, 25 µM GTP appears to extend the pause after the first insertion has occurred, and hence a greater fraction of the inserted mRNAs contains multiple insertions. These results also support the stuttering model in another way. That the frequency of insertions is similar in all respects at low CTP or UTP even though 9-fold less mRNA is made, is inconsistent with a posttranscriptional insertion process, since in that case the ratio of insertion activity (viral proteins) to mRNA would be 6 to 9-fold higher.

Stuttering implies that during the polymerase pause at the insertion site, the base pairing between the 3' end of the nascent mRNA and its template will be broken transiently, for slippage to occur. The precise number of these base pairs and their composition would then be important. We therefore examined the effect of incorporating base analogs into the mRNA chain which alter the strength of these interactions. Inosine (I) incorporation in place of guanosine will decrease the stability of the base pairing (and there must be at least one G:C pair available here for substitution), whereas 5-Br-uridine incorporation will have the opposite effect. Therefore, parallel transcription reactions were carried out under standard conditions, and under conditions in which 40% of the UTP was replaced with BrUTP, or 40% of the GTP with ITP.

When the frequency of G insertions in the mRNA was examined (Table 1), Br-U substitution for U was found to have little or no effect (a 1 G insertion frequency of 19.1% vs 24.4% for the parallel control, and a >1 G insertion frequency of 1.8% vs 1.7%). In contrast, inosine substitution for G increased the frequency of 1 G (or I) insertions to 38%, and the multiple G (or I) insertions were 10-fold more frequent now (19.8%). The incorporation of the analogs would have two effects; they would alter the stability of the intramolecular pairing upstream of the 3' end of the nascent chain, as well as that to the template at the 3' end. The absence of an effect when U is substituted with Br-U argues that the folding of the nascent chain has little effect on the insertion frequency. The effect of substituting inosine for G is then more likely to be due to the pairing with the template. In terms of the model in Figs. 2 and 3, this can be viewed as a decrease in the minimum time necessary for the initial slippage to occur, as well as subsequent slippages, because it would be easier to break the I:C pairs within the polymerase domain.

5.5 THE DIFFERENCE IN EDITING BETWEEN THE SEN/MV GROUP AND THE SV5/MUMPS GROUP

All attempts to show that the insertions can occur on preformed mRNAs were negative, and manipulations of the system *in vitro* can be interpreted in a coherent way for the predicted insertion site in terms of a stuttering model. The model can account for how the fraction of 1 G-inserted mRNAs is tightly controlled both *in vivo* (31% ± 2%), and *in vitro* under standard conditions (20% ± 2%, 4 determinations). Can it also offer any clues why 2 Gs are inserted at

Table I — mRNA synthesis *in vitro*

Reaction conditions	Total Colonies[a]	Colonies examined	No. Gs inserted at nt 1051-1053[b]			Total frequency of insertions
			0	1	>1	
Exp #1						
Mock	34	34	22 (65%)	9 (26%)	3 (9%)	35%
Standard	2688	224	174 (78%)	44 (18.5%)	6 (2.7%)	21.2%
25 µM CTP	408	170	135 (79%)	32 (18.8%)	3 (1.8%)	20.6%
25 µM UTP	362	151	112 (74%)	34 (22.5%)	5 (3.3%)	25.8%
			average + deviation:	19.9% ± 1.7%[c]	2.6% ± 0.5%[c]	22.5% ± 2.2%[c]
25 µM GTP	192	160	128 (80%)	19 (11.8%)	13 (8.1%)	19.9%
Exp #2						
Mock	10	0				
Standard	2688	176	130 (73.8%)	43 (24.4%)	3 (1.7%)	26.1%
40% Br UTP	952	167	133 (79.6%)	32 (19.1%)	3 (1.8%)	20.9%
40% ITP	317	176	74 (42%)	67 (38%)	35 (19.8%)	57.8%

[a] Neutral oligo[+] colonies.

[b] When 40% ITP is used, the inserted nucleotide is an I. Is are scored as Gs in the cDNA analysis performed.

[c] Average ± mean.

mRNAs were synthesized *in vitro* with purified Sendai virions, a 103 nt region containing the editing site was cloned, and resulting colonies were screened for the number of Gs at the insertion site by an oligonucleotide typing method (Vidal *et al.*, 1990b). Mock reactions lacked NTPs and contained an excess of EDTA over Mg^{++}, and probably measured a background of mRNAs which contaminate the virions.

high frequency in SV5 and mumps virus when insertions occur, rather than the 1 G inserted in Sendai and measles virus?

An alignment of the demonstrated (or suspected) insertion regions of 11 paramyxovirus templates is shown in Fig. 4. They all contain a minimun of 3 Cs on which the polymerase could pause and slip, but interestingly, the length of this C stretch is not well conserved; it varies from 3 to 5 Cs for the SEN/MV group, and from 4 to 7 Cs for the SV5/mumps group. In addition, there is little homology in these alignments downstream of the C stretch. However, the 1st, 3rd and 4th bases upstream of the C stretch are all uridines. The consensus sequences 3' UUUUCCC 5' (SEN/MV group) and 3' AAAUUCUCCCC 5' (SV5/mumps group) are evident. The meaning of the longer consensus sequence for the SV5/mumps group is unclear, and may only reflect the fact that the viruses in this group are more closely related to each other than those in the SEN/MV group.

A global consensus sequence 3' UUYUCCC 5' is also evident, in which the Y is a U in those viruses where 1 G is inserted, and a C where 2 Gs are inserted, and it is this difference which may be important. Fig. 2 shows a schematic representation of the insertion mechanism for each virus group. We do not know which template C is the pause site, or the number of base pairs between the nascent chain and the template. However, we would expect the latter to be limited to the minimum in this region, to facilitate the transient melting required for slippage. The model was originally demonstrated (Vidal *et al.*, 1990b) with 4 bp starting at the middle C of the sequence 3'UUYUCCC 5' to accommodate the proposed NDV editing sequence (3'UUCUUCCC 5' (Cattaneo *et al.*, 1989; Paterson *et al.*, 1989)) which is likely to be incorrect. The NDV sequence 3'UUUUUCCC 5' also exists nearby, and is more likely to be the correct one. The model in Fig. 2 uses the extended consensus sequence of each group, with 7 bp in each case and the polymerase paused on the penultimate C of the C stretch. There is presumably some pressure following the pause to displace the nascent chain upstream. For SV5/mumps, displacement upstream by 1 bp is unstable as it includes an A:C pair, whereas displacement by 2 bp creates a more stable intermediate. For the measles/Sendai group, on the other hand, an A:U rather than an A:C pair occurs at this position upon displacement by a single bp, and this intermediate is more stable than that obtained on a shift of 2 bp. Thus, the relative stabilities of the 1 and 2 bp misalignment intermediates for each virus could determine whether 1 or 2 Gs are inserted at high frequency when insertions occur. In the case of the SV5/mumps virus group, we note that once the initial 2 base slippage has occurred to avoid the unfavorable A:C pair, further rounds of slippage (when they occur) would now proceed one base pair at a time, as there are no further possibilities for forming A:C pairs. The distribution of G insertions in the mumps virus mRNAs has recently been determined (Paterson and Lamb, 1990; see Fig. 5) and fits nicely with this prediction.

A mechanism with many similar features has recently been proposed to account for the repeated TTGGGG sequences in tetrahymena telomers (Greider and Blackburn, 1989). These are added by a telomerase which contains an RNA template for this sequence within a larger chain. What is similar in this case is that the enzyme is proposed to pause on the template and then slip 6 nt upstream, and this distance is determined by the base pairing to the template.

<u>Paramyxovirus P Gene Editing Sites</u>

SEN/MV Group

```
                              *
SEN        g U U U U U U C C C          G U a U C C 5'

hPIV3      A U U U U U U C C C C C      u U u U C C 5'

bPIV3      A U U U U U U C C C C        u U c c u U 5'

(CDV)      A A U U U U U C C C C        G U G U C U 5'

MV         A A U U U U U C C C          G U G U C U 5'

(NDV)      g A U U U U U C C C          G g G U a C 5'

                              *
consensus  A U U U U U U C C C          G U G U C C 5'
           g A                          u g n c n U
```

SV5/Mumps Group

```
                              *
MUMPS      A A A U U C U C C C C C C    G g C c c U 5'

SV5        A A A U U C U C C C C        G u C c c U 5'

(PIV2)     A A A U U C U C C C C C C C u c g a u U 5'

(PIV4)     A A A U U C U C C C C C C    u u a u a U 5'

LPMV       A A A U U C U C C C C C C    G g C u g g 5'

                              *
consensus  A A A U U C U C C C C        G n C n n U 5'
                                        u   n     g
```

```
global
consensus         U U Y U C C C
```

Fig. 4 — Paramyxovirus P gene editing sites. The sequences of the various sites on the genomes or template strands are listed in the 3' to 5' direction, with the C stretch on which the insertions occur in the middle. Lower case letters indicate poor conservation, upper case letters indicate better conservation, and upper case and bold letters indicate complete conservation. The asterisk shows the pyrimidine residue which differentiates the SEN/MV and SV5/mumps groups: n refers to any base.

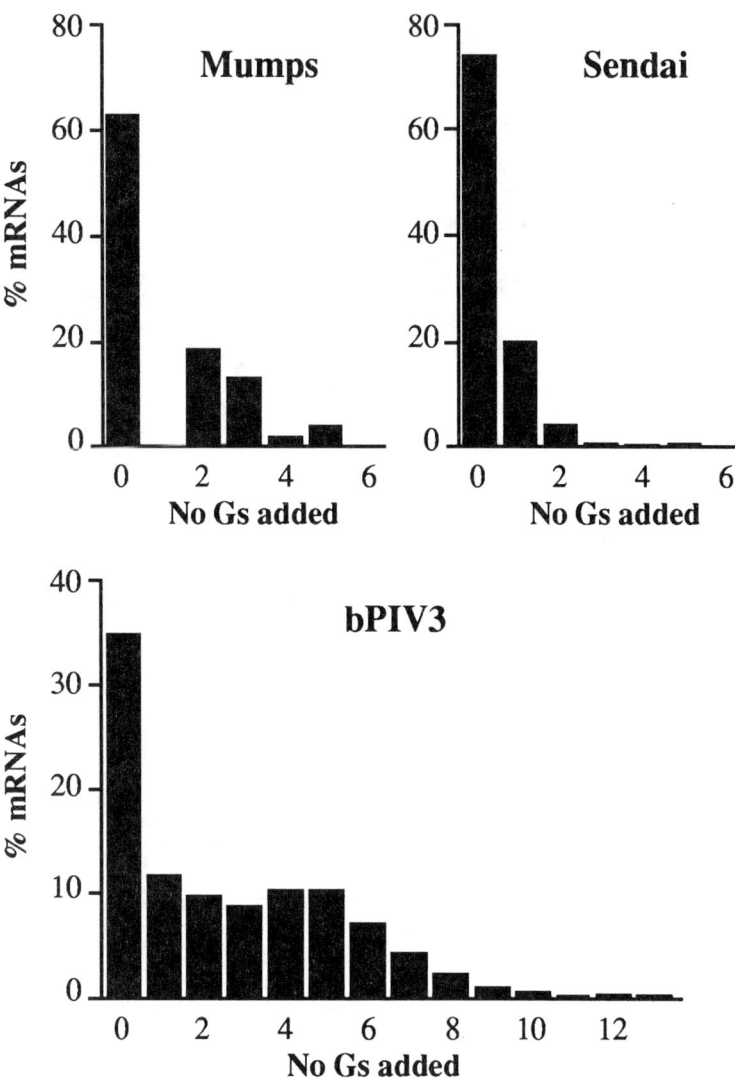

Fig. 5 — Distribution of G insertions in various paramyxoviruses. The distribution of the number of Gs inserted in the indicated paramyxovirus P gene mRNAs is shown as a bar graph. The mumps virus data are from Paterson and Lamb (1990), and were determined by sequencing 54 mRNA clones. The Sendai and bPIV3 data are from Pelet *et al.* (1991), and were determined by a primer extension method directly on the mRNA population. Zero Gs inserted refers to the unedited mRNA.

5.6 DIFFERENCES BETWEEN THE INITIAL AND SUBSEQUENT ROUNDS OF SLIPPAGE

Once the initial one or two base slippage/insertion has occurred, a second round is somehow avoided so that the 1 or 2 G insertions predominate. This could occur by the initial insertion(s) relieving the polymerase pause, in contrast to what would happen during polyadenylation. However, for Sendai virus, the initial insertion apparently fails to relieve the pause some of the times, as 7% of the mRNAs *in vivo* contain multiple G insertions, as do 2% of those made *in vitro*. When twelve >1 G inserted mRNAs were first examined by an oligonucleotide typing method (Vidal *et al.*, 1990a), those with the 2 Gs did dominate to some extent, but RNAs with 3 to 8 extra Gs were found in about equal amounts. A larger number of these mRNAs made *in vitro* has also been examined, and their distribution is not dissimilar (Vidal *et al.*, 1990b). What is striking here is that 12 Gs are as likely to be incorporated as 2 Gs when multiple insertions occur. This distribution suggests that in these cases the initial insertion has not only failed to relieve the pause, but may rather have increased it. However, when the whole distribution of G insertions was directly examined without a cloning step (by limited primer extension), a different pattern emerged. Two G insertions clearly predominated among the multiply-inserted mRNAs (Fig. 5). This is the pattern one would expect if the multiple G insertions occurred by simple repetition of the 1 G insertions (i.e., +2 G > +3 G > +4 G, etc.). These differing results make it difficult to know the precise distribution of the multiply-inserted mRNAs, but we suspect that the latter results (Fig. 5) are more accurate and that the cloning step used previously biased the distribution (by selecting larger insertions).

In any event, the question remains of how the 1 G insertions in Sendai virus predominate when insertions occur. Since a U:G base pair will be formed between the nascent mRNA and the template upon this insertion, and the polymerase is back to where it started before the insertion, it should be even easier to slip again. This is highlighted by the finding that low GTP concentrations and substituting inosines for guanosines in the nascent chain (*in vitro*) have much stronger effects on the multiple G insertions than on the 1 G insertions (Table 1).

5.7 BOVINE PARAINFLUENZA VIRUS TYPE 3, THE EXCEPTION THAT STRENGTHENS THE RULE

A remarkable feature of the paramyxovirus editing mechanism is that for each virus group, the requisite number of Gs are added at high frequency to provide access to the alternate downstream ORF, i.e. 2 Gs are added in the SV5/mumps group to go from V to P and mRNAs with only 1 G inserted are very rare, whereas +1 G is the predominant event for the SEN/MV group to go from P to V. This is best illustrated by comparing Figs 1 and 5.

PIV3 (parainfluenza virus type 3) is closely related to SEN and MV, but is slightly more complicated. The P gene sequences of both the human (h) (Spriggs and Collins, 1986; Galinski *et al.*, 1986; Luk *et al.*, 1986) and the bovine (b)

(Sakai *et al.*, 1987) viruses are known, and these contain one more ORF (called D). The D ORF follows shortly after the C ORF, in the same reading frame (Fig. 1). On the basis of sequence comparisons, Cattaneo *et al.* (1989) suggested two editing sites for hPIV3, both of which would favor single base slippages according to the above rule. 1 G insertions at the first site (site a) cannot produce a functional V mRNA for hPIV3, as there are several stop codons between this site and the cys-rich domain. However, a 2 G insertion here would give rise to an mRNA which has fused the N-terminus of P to 131 amino acids of the D ORF. A second insertion site (site b), ca. 300 nt downstream and at the same relative location as the site which operates in MV and SEN, lies downstream of the stop codons and can produce a functional V mRNA. hPIV3 would then appear to use two insertion sites; the first to express the D ORF exclusively, and the second for V.

The P gene of bPIV3 is organized identically to that of hPIV3 and contains the same two putative insertion sites (Fig. 1). However, there are no stop codons between the first site and the cys-rich domain of the V ORF. This creates the very unusual situation where a block of 320 nt can potentially be expressed in all 3 reading frames. As it is unlikely that such a situation would exist by chance, this suggests that for bPIV3 the first editing site will be used to express both the V and D proteins (Fig. 1). Moreover, if the second site also operates in bPIV3, there could potentially be 3 different forms of V as well as multiple forms of P.

A primer extension method was used to quantitate the fraction of mRNAs with insertions, and the distribution of the number of bases inserted (Pelet *et al.*, 1991). Examination of the bPIV3 mRNAs indicated that site b was non-functional, whereas site a operates frequently. Moreover, the distribution of G insertions at site a was remarkably broad; from 1 to 6 Gs were added with virtually equal frequency (Fig. 5). Insertions at site a can then give rise to both V and D protein mRNAs, because a very broad distribution of Gs are added when insertions occur. This broad distribution is very different from the editing sites of Sendai virus or SV5, where predominantly one form of edited mRNA (containing either a 1 or 2 G insertion, respectively) is created, to access the single overlapping ORF of these latter viruses. Although SEN, MV, and bPIV3 all insert Gs within the common sequence 5'AAAAAGGG, bPIV3 does this somewhat differently, and there is a plausible mechanism for how this could occur in the stuttering model. In this model, the polymerase would pause after incorporating the second G at the editing site (top line, Fig. 6). In some cases, the pause would be too short for the base pairing between the 3' end of the nascent mRNA and the template to be broken. Here the polymerase would continue on, without creating insertions. When the pause is long enough, the base pairing is transiently interrupted, and there is some pressure to move the 3' end of the mRNA upstream. For bPIV3 the pause would presumably be longer than for SEN, as 65% of the bPIV3 mRNAs contain inserted Gs versus 25% for SEN. However, once the initial round of slippage/insertion(s) has occurred (second and third lines, Fig. 6), the polymerase is back to the same position as before, except that one G:U pair has been formed in the misalignment. A second round of slippage is then favored if nothing else has changed, since it should be easier to break the

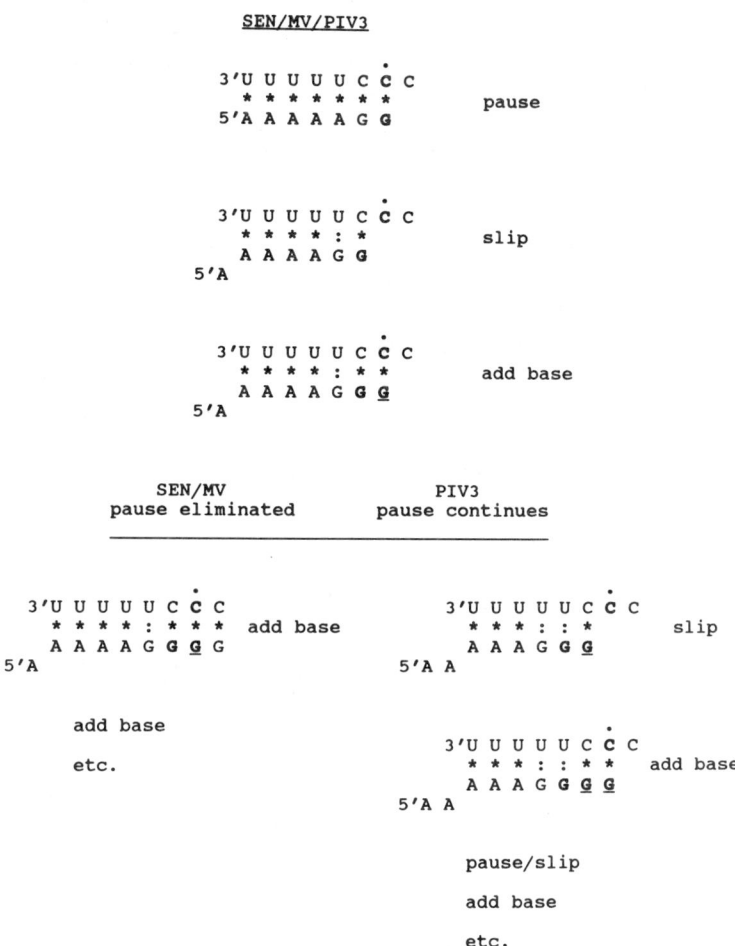

Fig. 6 — A model to explain the different frequencies of multiple G insertions for SEN/MV and PIV3. The polymerase pauses after adding the second G to the mRNA (in bold, bottom line of top duplex), and the mRNA slips one base upstream, creating a single G:U pair (double dots). The length of the pause determines the total fraction of mRNAs with insertions, and presumably the pause is stronger for bPIV3 than for SEN. Addition of the pseudo-templated G brings the polymerase back to the pause site (the template C with dot above), creating a single base insertion (underlined). For SEN/MV, the pause is presumably eliminated at this point, the next truly templated G is added, and the polymerase continues in a strictly processive mode. For PIV3, the pause is not eliminated. Because the G:U pair makes it easier to break the bonds between the mRNA and the template for slippage to reoccur, multiple rounds of slippage and insertion are favored, and a large majority of the mRNAs with insertions contain multiple insertions. After about 6 rounds, the probability that the pause is eliminated eventually increases, or the process ceases as oligo G formation becomes self-limiting.

new base pairs for slippage to recur. Nevertheless, both *in vivo* and *in vitro*, 80% or more of SEN mRNAs which contain insertions have inserted only a single G. For 1 G insertions to predominate, something else must have changed when the polymerase returns to the pause site after the initial events, to prevent a second round. For example, protein modification of the polymerase or template could have occurred during the first round, which eliminates the pause. It is apparently this something else which has failed to occur in large part during bPIV3 mRNA editing, because here insertions of >1 G are now more than 5 times as frequent as those with only 1 G.

Somewhat similar situations can be created with SEN *in vitro*, by altering reaction conditions which are thought either to extend the pause, or to decrease the time required for slippage to occur (Vidal *et al.*, 1990b). For example, as discussed above, lowering the GTP concentration during mRNA synthesis does not appear to extend the initial pause at the insertion site, but does appear to extend the subsequent pauses. This suggests that the initial and subsequent pauses operate somewhat differently. Moreover, partial replacement of GTP with ITP during synthesis only doubles the frequency of mRNAs with 1 G (or I) insertions, whereas those with >1 G (or I) insertions increase 10-fold.

5.8 CONCLUSIONS

Paramyxovirus mRNA editing, like ribosomal frameshifting of retroviruses (Jacks and Varmus, 1985) or coronaviruses (Brierley *et al.*, 1987, 1989), appears to be composed of two parts: a pause, and a slippery sequence which allows for alternate base pairing (Brierley *et al.*, 1989). The slippery sequence for paramyxoviruses (3' UUYUCCC) is fairly clear. However, the counterpart to the downstream template structure which causes ribosomal pausing is difficult to even guess at. DNA dependent RNA polymerases have long been known to pause in the middle of genes (Chamberlin, 1976; Von Hippel *et al.*, 1984; Platt, 1986), and although there is some evidence that this is due to an intrinsic property of the template (Reines *et al.*, 1987), its nature is equally unclear. Further progress in this form of mRNA editing will require some understanding of the nature of the pause. Moreover, the reasons why some viruses have chosen to express alternate downstream ORFs via ribosomal frameshifting, whereas others have chosen "polymerase frameshifting", remains obscure.

The G insertions do not represent mRNA editing in a strict sense, as the mRNA has not been altered <u>after</u> its synthesis. However, it is a form of editing in that it results in a specific change of the mRNA relative to its template which is important for translation. Compared with the "true" forms of editing which take place after synthesis of the transcript, the stuttering mechanism is clearly different.

REFERENCES

Banerjee, A.K. (1987) Transcription and replication of rhabdoviruses. *Microbiol. Rev.* **51** 66-87.

Berg, M., Hjertner, B., MorenoLopez, J. & Linne, T. (1992) The P gene of the porcine paramyxovirus LPMV encodes three possible polypeptides P, V and C: the P protein mRNA is edited. *J. Gen. Virology* **73** 1195-1200.

Brierley, I., Boursnell, M.E.G., Binns, M.M., Bilimoria, B., Blok, V.C., Brown, T.D.K. & Inglis, S.C. (1987) An efficient ribosomal frameshifting signal in the polymerase encoding region of the coronavirus IBV. *EMBO J.* **6** 3779-3785.

Brierley, I., Digard, P. & Inglis, S.C. (1989) Characterization of an efficient Coronavirus ribosomal frameshifting signal: requirement for an RNA pseudoknot. *Cell* **57** 537-547.

Cattaneo, R., Kaelin, K., Baczko, K. & Billeter, M.A. (1989) Measles virus editing provides an additional cysteinerich protein. *Cell* **56** 759-764.

Chamberlin, M.J. (1976) In Chamberlin, M.J. and R. Losick (eds). RNA Polymerase. Cold Spring Harbor Laboratory Press. Cold Spring Harbor, NY, pp. 159-192.

Curran, J. & Kolakofsky, D. (1988) Ribosomal initiation from an ACG codon in the Sendai virus P/C mRNA. *EMBO J.* **7** 245-251.

Curran, J. & Kolakofsky, D. (1989) Scanning independent ribosomal initiation of the Sendai virus Y proteins *in vitro* and *in vivo*. *EMBO J.* **8** 521-526.

Curran, J., Boeck, R. & Kolakofsky, D. (1991) The Sendai virus P gene expresses both an essential protein and an inhibitor of RNA synthesis by shuffling modules via mRNA editing. *EMBO J.* **10** 3079-3085.

Galinski, M.S., Mink, M.A., Lambert, D.M., Wechsler, S.L. & Pons, M.W. (1986) Molecular cloning and sequence analysis of the human parainfluenza 3 virus mRNA encoding the P and C proteins. *Virology* **155** 46-60.

Galinski, M.S., Troy, R.M. & Banerjee, A.K. (1992) RNA editing in the phosphoprotein gene of the human parainfluenza virus type 3. *Virology* **186** 543-550.

Giorgi, C., Blumberg, B.M. & Kolakofsky, D. (1983) Sendai virus contains overlapping genes expressed from a single mRNA. *Cell* **35** 829-836.

Greider, C.W. & Blackburn, E.H. (1989) A telomeric sequence in the RNA of Tetrahymena telomerase required for telomere repeat synthesis. *Nature* **337** 331-336.

Gupta, M. & Kingsbury, D. (1984) Complete sequences of the intergenic and mRNA start signals in the Sendai virus genome: homologies with the genome of vesicular stomatitis virus. *Nucleic Acids Res.* **12** 3829-3841.

Herrler, G. & Compans, R.W. (1982) Synthesis of mumps virus polypeptides in infected vero cells. *Virology* **119** 430-438.

Iverson, L.E. & Rose, J.K. (1981) Localized attenuation and discontinuous synthesis during Vesicular Stomatitis Virus transcription. *Cell* **23** 477-484.

Jacks, T. & Varmus, H.E. (1985) Expression of the Rous sarcoma virus pol gene by ribosomal frameshifting. *Science* **230** 1237-1242.

Kolakofsky, D. & Roux, L. (1987) The molecular biology of paramyxoviruses. In Bercoff P. (ed.) The Molecular Basis of Viral Replication. Plenum Publishing, New York, pp. 277-297.

Krakow, J.S., Rhodes, G. & Jovin, T.M. (1976) In Chamberlin M.J. and R.

Losick, (eds). RNA Polymerase Cold Spring Harbor Laboratory Press. Cold Spring Harbor, NY, pp. 127-158.

Luk, D., Sanchez, A. & Banerjee, A.K. (1986) Messenger RNA encoding the phosphoprotein (P) gene of human parainfluenza virus 3 is bicistronic. *Virology* **153** 318-325.

McGeoch, D.J. (1979) Structure of the gene N: gene NS intercistronic junction in the genome of vesicular stomatitis virus. *Cell* **17** 673-681.

Paterson, R.G., Thomas, S.M. & Lamb, R.A. (1989) Specific nontemplated nucleotide addition to a simian virus 5 mRNA: prediction of a common mechanism by which unrecognized hybrid P-cystein rich proteins are encoded by paramyxovirus "P" genes. In Kolakofsky, D. & B.W.J. Mahy (eds). Genetics and Pathogenicity of Negative Stranded Viruses. Elsevier/North-Holland Publishing Co., Amsterdam, pp. 232-245.

Paterson, R.G. & Lamb, R.A. (1990) RNA editing by G-nucleotide insertion in mumps virus "P" gene mRNA transcripts. *J. Virol.* **64** 4137-4145.

Pelet, T., Curran, J. & Kolakofsky, D. (1991) The P gene of bovine parainfluenza virus 3 expresses all three reading frames from a single mRNA editing site. *EMBO J.* **10** 443-448.

Platt, T. (1986) Transcription termination and the regulation of gene expression. *Annu. Rev. Biochem.* **55** 339-372.

Reines D., Wells, D., Chamberlin, M.J. & Kane, C.M. (1987) Identification of intrinsic termination sites *in vitro* for RNA polymerase II within eukaryotic gene sequences. *J. Mol. Biol.* **196** 299-312.

Robertson, J.S., Schubert, J.S. & Lazzarini, R.A. (1981) Polyadenylation sites for influence virus mRNA. *J. Virol.* **38** 157-163.

Ruterhouser, E.C. & Richardson, J.P. (1989) Identification and characterization of transcription termination sites in the *Escherichia coli* lacZ gene. *J. Mol. Biol.* **208** 2343.

Sakai, Y., Suzu, S., Shioda, T. & Shibuta, H. (1987) Nucleotide sequence of the bovine parainfluenza 3 virus genome: its 3' end and the genes of NP, P, C and M proteins. *Nucleic Acids Res.* **15** 2927-2944.

Schubert, M., Keene, J.D., Herman, R.C., & Lazzarini, R.A. (1980) Site on the vesicular stomatitis virus genome specifying polyadenylation and the end of the L gene mRNA. *J. Virol.* **34** 550-559.

Spriggs, M.K. & Collins, P.L. (1986) Sequence analysis of the P and C protein genes of human parainfluenza virus type 3: patterns of amino acid sequence homology among paramyxovirus proteins. *J. Gen. Virol.* **67** 2705-2719.

Thomas, S.M., Lamb, R.A. & Paterson, R.G. (1988) Two mRNAs that differ by two nontemplated nucleotides encode the amlno coterminal proteins P and V of the paramyxovirus SV5. *Cell* **54** 891-92.

Von Hippel, P.H., Bear, D.G., Morgan, W.D. & McSwiggen, J.A. (1984) Protein-nucleic acid interactions in transcription: a molecular analysis. *Ann. Rev. Biochem.* **53** 389-446

Vidal, S., Curran, J. & Kolakofsky, D. (1990a) Editing of the Sendai virus P/C mRNA by G insertion occurs during mRNA synthesis via a virus-encoded activity. *J. Virol.* **64** 239-246.

Vidal, S., Curran, J. & Kolakofsky, D. (1990b) A Stuttering model for paramyxovirus P mRNA editing. *EMBO J.* **9** 2017-2022.

6

EDITING OF MAMMALIAN APOLIPOPROTEIN B mRNA BY SITE-SPECIFIC RNA DEAMINATION

Peter Hodges[+] and James Scott[*]
[*]The MRC Molecular Medicine Group, Department of Medicine, Royal Postgraduate Medical School, Du Cane Road, London W12 0NN, UK
[+]Institute of Cell and Molecular Biology, University of Edinburgh, King's Buildings, Mayfield Road, Edinburgh EH9 3JR, UK

Cytidine 6666 of the 14 kilobase mRNA encoding apolipoprotein B (apo B) is deaminated by an RNA editing activity. This deamination creates a uridine base, changing the 2153rd codon from CAA (encoding glutamine) to a UAA stop codon. The truncation of the apo B protein removes a domain required for receptor interaction and causes lipoprotein particles to be rerouted through the body. By regulating the RNA editing activity, the balance between the two apo B proteins is tissue specific, developmentally coordinated, and hormonally and nutritionally responsive.

The RNA editing activity is a protein which catalyzes a site-specific deamination with enzymatic kinetics. The specificity of editing is provided by binding to a target sequence in the RNA downstream of the editing site, which directs the deamination of the cytidine or cytidines at a fixed distance upstream. The rapid progress in elucidating the apo B mRNA editing process has come from the development of an *in vitro* editing reaction and a sensitive and specific assay for editing, the partial purification of the deaminase, and the combination of *in vitro*, cell culture and animal studies. Several research groups are keenly pursuing the purification of the editing activity, the isolation of cDNAs encoding the enzyme's components by biochemical and genetic approaches, and the identification of additional mRNAs edited by this deaminase activity.

6.1 THE STRUCTURE AND FUNCTION OF APOLIPOPROTEIN B

In mammals two closely-related forms of apo B circulate in the blood. The smaller form of apo B (241 kDa) is designated apo B48 because it is 48% of the size of the larger apo B100 (512 kDa) on SDS polyacrylamide gel electrophoresis. Apo B48 is made in small intestinal enterocytes, where it is necessary for the transport of dietary fat as triglyceride-rich lipoprotein particles called chylomicrons. Triglyceride is removed from the particle by the extracellular action of endothelial cell lipases throughout the body, but the entire particle is taken up by only liver cells, which clear the depleted chylomicron remnants.

Apo B synthesized in the liver is responsible for the secretion of endogenously synthesized lipid and repackaged dietary lipid. In humans the liver makes solely apo B100, whereas in rodents apo B48 and apo B100 are both synthesized in the liver. Apo B100 contains additional carboxy-terminal sequences which form the binding site for the low-density lipoprotein (LDL) receptor, and can target the particle for endocytosis throughout the body. Since circulating levels of LDL are strongly correlated with the risk of heart disease, the regulation of apo B synthesis is a topic of medical as well as biological importance.

Genetic and molecular biological studies have demonstrated that in mammals there is a single copy of the apo B gene (Fig. 1). Coincident with the structural characterization of the apo B gene and the sequencing of the 14,121 bp cDNA that encodes for apo B100, the mapping of epitopes for monoclonal antibodies and sequence specific polyclonal antibodies demonstrated that the carboxy terminal of apo B48 was in the middle of the huge (7572 base pair) exon 26 of the apo B gene (Chen *et al.*, 1987; Powell *et al.*, 1987). Codon 2153 in the apo B gene and the apo B100 cDNA is CAA for glutamine. In the apo B48 mRNA, this codon is UAA, a stop translation codon (Chen *et al.*, 1987; Powell *et al.*, 1987; Hospattankar *et al.*, 1987; Higuchi *et al.*, 1988; Aburatani *et al.*, 1989). Differential splicing of apo B100 and apo B48-specific exons, and more unusual mechanisms such as somatic mutation in cells that make apo B48, were excluded as the molecular basis for the difference between apo B100 and apo B48 mRNAs. It was therefore proposed that in cells that make apo B48 the mRNA undergoes an editing reaction in which cytidine is replaced by uracil or by a base which is read as uracil by reverse transcriptase and the translation machinery. Amino acid sequencing and the mapping of site-specific monoclonal antibodies has confirmed that isoleucine 2152 represents the carboxy terminal of plasma apo B48. Apo B48 therefore consists of the amino terminal 1-2152 amino acid residues of apo B100 and is fully competent for lipoprotein assembly and secretion. It lacks the carboxy terminal domain found in apo B100 that mediates the clearance of LDL from the circulation by the LDL receptor pathway.

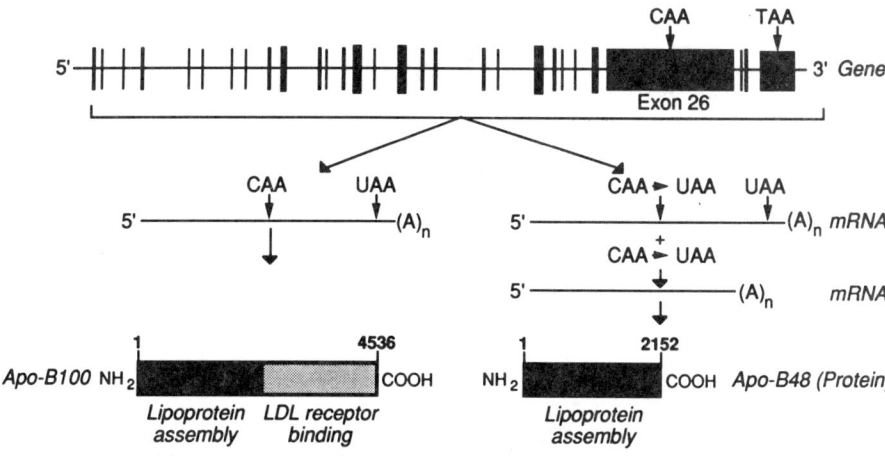

Fig. 1 — The intron-exon organization of the apo B gene and the origins of apo B100 and apo B48 mRNAs and proteins.

6.2 METHODS IN THE STUDY OF APO B mRNA EDITING

The initial discovery of apo B mRNA editing came from sequencing clones isolated from cDNA libraries. Nearly all cDNAs isolated from human or rabbit intestinal libraries contain TAA at codon 2153 (Powell *et al.*, 1987; Chen *et al.*, 1987; Higuchi *et al.*, 1988; Aburatani *et al.*, 1989). Similarly, TAA containing clones are present in liver cDNA libraries, albeit at a much lower level (Higuchi *et al.*, 1988). Polymerase chain reaction (PCR) amplification of cDNA circumvents the tedium of creating cDNA libraries. Edited and unedited cDNAs can be distinguished by differential hybridization with oligonucleotides specific for CAA or TAA, either in total PCR products or as individual cDNA clones. Hybridization to a large number of PCR-generated cDNA clones is one of the most accurate methods of quantitating apo B mRNA editing (Wu *et al.*, 1990).

An alternative assay was developed by Driscoll *et al.* (1989), based on primer extension (Fig. 2). An oligonucleotide is hybridized to (un)edited mRNA or PCR- amplified cDNA just downstream of the editing site, and is extended by reverse transcriptase in the presence of dideoxy GTP (and dATP, dCTP, and dTTP). While reverse transcriptase terminates at the unedited C, editing allows the polymerase to read through the site and terminate opposite the next C. The primer extension assay is sensitive and specific, and, more importantly, speeds the analysis of multiple samples. This assay was crucial to the development of an *in vitro* editing reaction.

Several additional assays have been described. Differential hybridization with oligonucleotides directly on Northern blots has the advantage of distinguishing

Primer extension

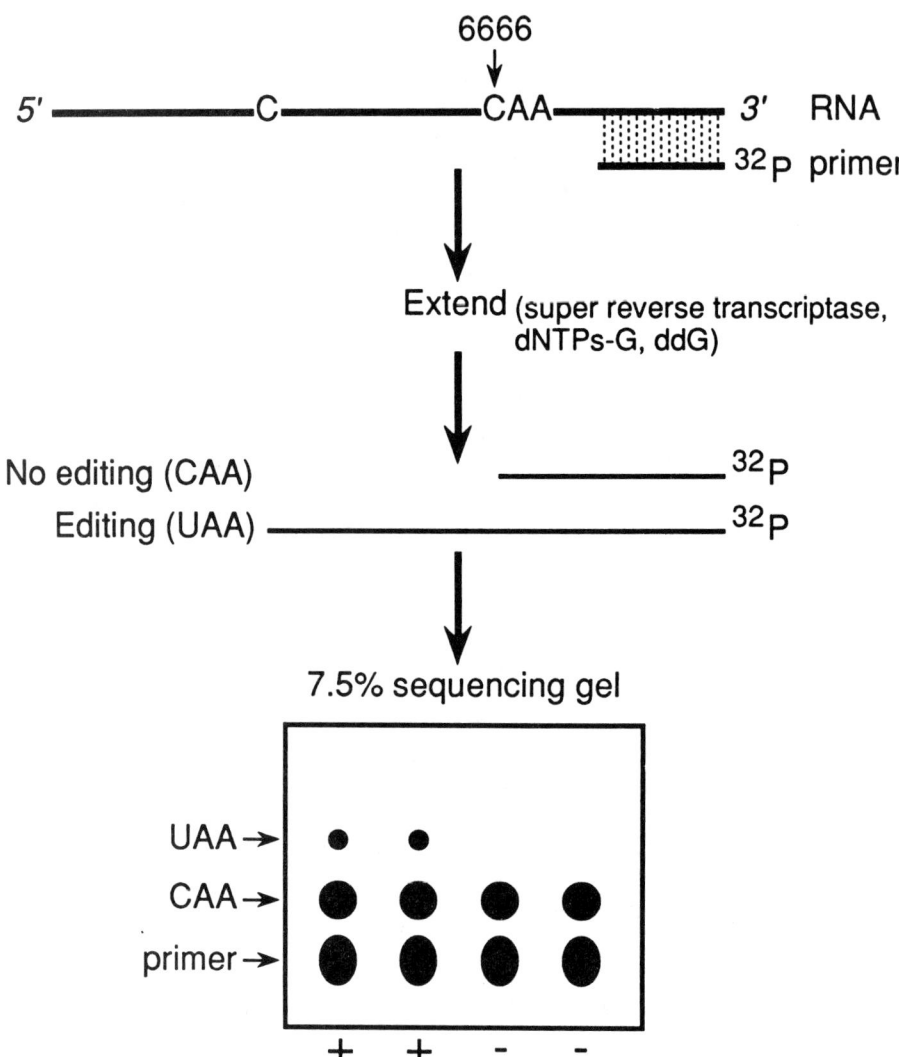

Fig. 2 — Primer extension assay used to demonstrate edited and unedited transcripts.

full length from prematurely polyadenylated RNA species (see below). Direct sequencing of PCR amplified cDNA can be quantitated by densitometry of the C and T lanes at the editing site. Taking advantage of the sensitivity of PCR to proper annealing of the 3' nucleotide of the primer, a "competimer" PCR reaction has been developed. The sensitivity of single base mismatches to detection with RNAse A has been exploited by another group to develop an RNAse protection assay to detect editing. To be useful, an assay must be quantitative, accurate at lower levels of editing and with little input edited RNA, fast, and easy to perform simultaneously on multiple samples. The primer extension assay has remained the gold standard.

A breakthrough in the investigation of apo B mRNA editing was the development of an *in vitro* editing reaction (Driscoll *et al.*, 1989). Exogenously added RNA can be edited *in vitro*, posttranscriptionally, by a cytoplasmic S100 extract from cell lines or animal tissues that edit endogenous RNA (McArdle 7777 rat hepatoma cells, rat liver, rabbit, rat, and baboon intestine), but not extracts from cells or tissues with little or no *in vivo* editing activity (HepG2 human hepatoblastoma cells, HeLa human cervical epidermoid carcinoma cells, or baboon liver, spleen, kidney, testes, or adrenal gland tissue). An *in vitro* editing reaction and soluble editing extracts have allowed the definition of the RNA target site that directs editing, definition of the reaction mechanism, and characterization and partial purification of the editing deaminase.

6.3 NUCLEAR RNA EDITING AND THE CONNECTION TO POLYADENYLATION

The primary source of editing activity has been cytoplasmic S100 extracts, but these extracts typically contain nuclear components. Further, nuclear extracts are functional in coupled *in vitro* transcription and editing (Chen *et al.*, 1990) and editing can influence the nuclear events of 3' end formation and polyadenylation (see below). Lau *et al.* (1991) have accumulated convincing evidence that the activity is nuclear. They showed that nuclei have a higher specific activity of the editing enzyme, although a greater quantity of the activity is recovered in the "cytoplasmic" fraction. After fractionating nascent and mature RNA from rat liver, the unspliced unpolyadenylated nuclear pre-mRNA is largely unedited (2%), spliced but unpolyadenylated mRNA is more highly edited (8%), unspliced but polyadenylated pre-mRNA is more completely edited (25%), and the spliced and polyadenylated mRNA still in the nucleus is as completely edited as in the cytoplasm (about 50%). It can be concluded that editing is a posttranscriptional nuclear function, coincident with splicing and polyadenylation.

In humans and rabbits, apo B mRNA editing allows premature cleavage and polyadenylation within exon 26 to truncate the 14 kb mRNA into a 7-8 kb species. As deduced from cDNA sequences, cryptic polyadenylation occurs at one major site and four minor sites (Table 1). Consensus or variant polyadenylation signal sequences can be identified upstream of these sites; the major site follows an AAUUAA variant sequence. This major signal has mutated in the rat and mouse genes, which do not produce shortened mRNAs. Cryptic polyadenylation is specific for the edited mRNA. Premature polyadenylation of

Table 1 — Premature polyadenylation sites

| Species | Signal | | PolyA addition site | | Effect on |
	Sequence	Spacing	Sequence	Position	unedited mRNA
Human	AAUUAA	19 bases	UA:U	6775	Termination
	ACCAAA	10 bases	CA:G	6910	Polylysine
	AAUAAA	20 bases	GUC A:AA	7080	Polylysine
	UAUAAA	25 bases	GCU:GAG	7121	Polylysine
Rabbit	AAUUAA	19 bases	UA:U	6775	Termination
	AAUUAA	16 bases	UUA:UAU	6815	Polylysine
	AAUAAA	20 bases	UUC A:AA	7080	Polylysine

the unedited RNA has not been seen; a CAA-specific oligonucleotide does not hybridize to the shorter mRNA on Northern blots and cells which do not edit the apo B mRNA do not produce shortened transcripts. Can the polyadenylation apparatus sense a termination codon in the nucleus, before translation? Alternatively, do the editing and polyadenylation activities cooperate to act together on the RNA?

A connection between RNA editing and polyadenylation is a tantalizing notion, but it has not yet been distinguished from the consistent use of the cryptic site and instability of incorrectly polyadenylated mRNAs. Premature polyadenylation of unedited mRNAs at any of the minor sites identified by Powell *et al.* (1987) would cause the uninterrupted translation of the polyadenylate into polylysine, which may destabilize the mRNA. However, polyadenylation at the primary site without editing would change tyrosine codon 2189 (UAU) to a UAA stop codon, giving apo B48 protein elongated by 36 amino acids. A precedent exists for an mRNA with no 3' untranslated region and a stop codon changed by polyadenylation from UAG to UAA (Jenh *et al.*, 1986), and the termination codons of several mitochondrial transcripts are created by cleavage and polyadenylation (Ojala *et al.*, 1981). Thus, premature polyadenylation could serve the same function as RNA editing, to truncate the apo B coding region. This may explain the findings of Yao *et al.* (1992), who find truncated apo B48-like protein produced from transfected genes even when a mutation was introduced (CAA to GAT) to prevent the creation of a stop codon and no editing is detected in cDNA sequences. Unfortunately, premature polyadenylation was not studied in this report. It is possible that, with no mutations in the actual editing enzyme recognition sequence (see below), the editing enzyme binds to the apo B mRNA, is unable to edit the sequence, but still attracts the polyadenylation machinery to use the cryptic site.

Polyadenylation within exon 26 may be related to two other unique aspects of this region. The 150 bases surrounding the editing site have an unusually high content of A and U, a characteristic more common of 3' untranslated regions. The editing site is found in a huge exon (7572 bp). Exons are normally confined to a narrow size range (150-200 bp) except for untranslated regions and 3' terminal exon coding sequences (Hawkins, 1988; Naora and Deacon, 1982). Perhaps the characteristics of an untranslated region and the lack of nearby donor and acceptor splice sites change the pathway of RNA processing for this exon, providing access for an RNA editing enzyme. Berget and coworkers have proposed that exon size is an important determinant in pre-mRNA processing; the fixed exon size promotes cooperation between the upstream and downstream splice sites in "exon definition" (Robberson *et al.*, 1990). Further, it is proposed that 3' terminal exons escape this requirement by interaction of the polyadenylation apparatus with splicing factors. The RNA editing apparatus may also cooperate in exon definition, interacting with the splicing or polyadenylation machinery to redefine the exon.

The correlation between premature polyadenylation and editing has been reproduced with a transfected minigene containing only 354 bases of apo B sequence (Böstrom *et al.*, 1989). It will be interesting to see whether mutation of the polyadenylation site affects editing, whether premature polyadenylation requires that editing creates a termination codon (rather than a CUA to UUA change for instance), and whether nearby splicing sites can interfere with cryptic polyadenylation.

6.4 DEFINITION OF THE mRNA TARGET SITE

What directs an RNA editing activity to modify a specific cytidine in the 14,000 base apoB mRNA? A first clue came from cross-species sequence comparisons in the vicinity of the edited C (Davies *et al.*, 1989). As more mammalian species have been studied, it has become clear that a short stretch of sequence surrounding the editing site is conserved more highly than other regions of the mRNA (Teng *et al.*, 1990a). As shown in Fig. 3, a 23 base sequence is invariant among the six species. Over a 37 nucleotide span only three changes are seen, each unique to one species, giving an average pairwise homology of 97.3%. This must be significant, since selection to maintain the amino acid sequence through this sequence would still allow changes at the third base of codons, yet none are seen. Indeed, the three nucleotide differences all change the amino acid encoded. A functional role for this RNA sequence has been demonstrated: placed into a foreign RNA context, as little as 26 bases can direct RNA editing, and mutagenesis shows that a core downstream element is most important in targeting the editing activity.

The apo B sequence can be transferred into other mRNAs and retains its ability to be edited, *in vivo* and *in vitro*. When transfected into tissue culture cells, apo B sequences as short as 354 nucleotides (Böstrom *et al.*, 1989) or 26 nucleotides (Davies *et al.*, 1989) could be edited, although Böstrom *et al.* (1989) could not detect the editing of a 63 nucleotide sequence. *In vitro*, as little as 260 nucleotides (Chen *et al.*, 1990) or 55 nucleotides (Driscoll *et al.*, 1989) of apo B

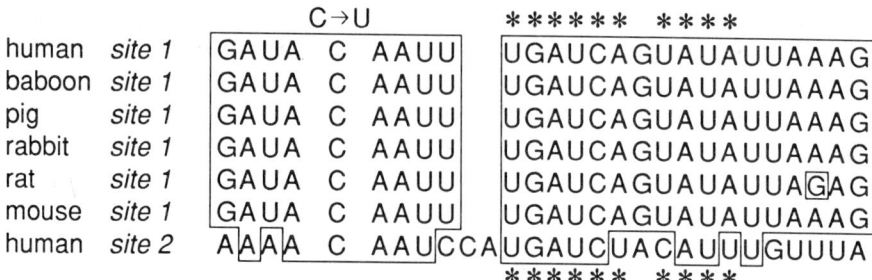

Fig. 3 — Sequence conservation in the apo B mRNA editing site. The editing primary site is shown in a number of species, and the secondary site shown for humans. The asterisks indicate the nucleotides essential to editing, as determined by site-directed mutagenesis (see Fig. 4).

sequence directed efficient editing. Although 26 nucleotides of apo B mRNA in a 73 nucleotide RNA was not initially found to be edited in crude S100 extracts from McArdle 7777 cells (Driscoll *et al.*, 1989), in an optimized assay with extract from rat enterocytes this RNA is edited with about 25% of the efficiency of the 55-nucleotide and larger RNAs (Shah *et al.*, 1991). Similar results were obtained by Backus and Smith (1991). Therefore, a highly conserved core sequence of 26 nucleotides is sufficient to confer editing specificity when introduced into a novel RNA context.

Site-directed mutagenesis of the editing site has been used to define the sequence requirements more precisely (Chen *et al.*, 1990; Shah *et al.*, 1991). Shah *et al.* have dissected the 55-nucleotide sequence that undergoes efficient editing *in vitro*. A series of scanning mutants were generated where 6-nucleotide blocks of the 55-nucleotide region were changed to anti-sense sequence. As shown in Fig. 4, the results point to a critical region downstream of the editing site. Mutation of the 12-nucleotide region immediately downstream of C-6666 abolished editing, and mutation of the two blocks flanking this 12-nucleotide region significantly reduced editing. To analyze this region more finely, a panel of 46 single-base substitutions or deletions was generated in the region between nucleotide positions -9 and +19. An 11-nucleotide sequence downstream of the edited C (UGAUCAGUAUA; position +5 to +15) was identified where almost all point mutations abolished or greatly reduced editing *in vitro*. Shah *et al.* (1991) proposed that this sequence is the binding site for the apo B editing enzyme, and that binding to this sequence provides the specificity to modify the upstream cytidine. No obvious RNA secondary structure can be predicted for this sequence which is consistent with the mutational data, so the sequence may be recognized in a disordered structure.

Chen *et al.* (1990) showed that mutations in the sequences surrounding the editing site led to little or no change in editing activity or even to increased efficiency. This was also seen by Shah *et al.* (1991), and confirmed by the demonstration by Backus and Smith (1991) that the downstream specificity element can be placed into a foreign context to induce the editing of an upstream C. However, the immediate context of the edited C does modulate the efficiency of editing. Spacing is most crucial; the efficiency of editing is very intolerant of deletions or insertions. Some flexibility exists, as other cytidines introduced in the vicinity of the edited C can be edited too, either independently or in concert with the natural site. These effects can be demonstrated by a series of mutations around the editing site studied by Chen *et al.* (1990) (Table 2). Moving the editing site one nucleotide closer to the downstream target site (mutation -A)

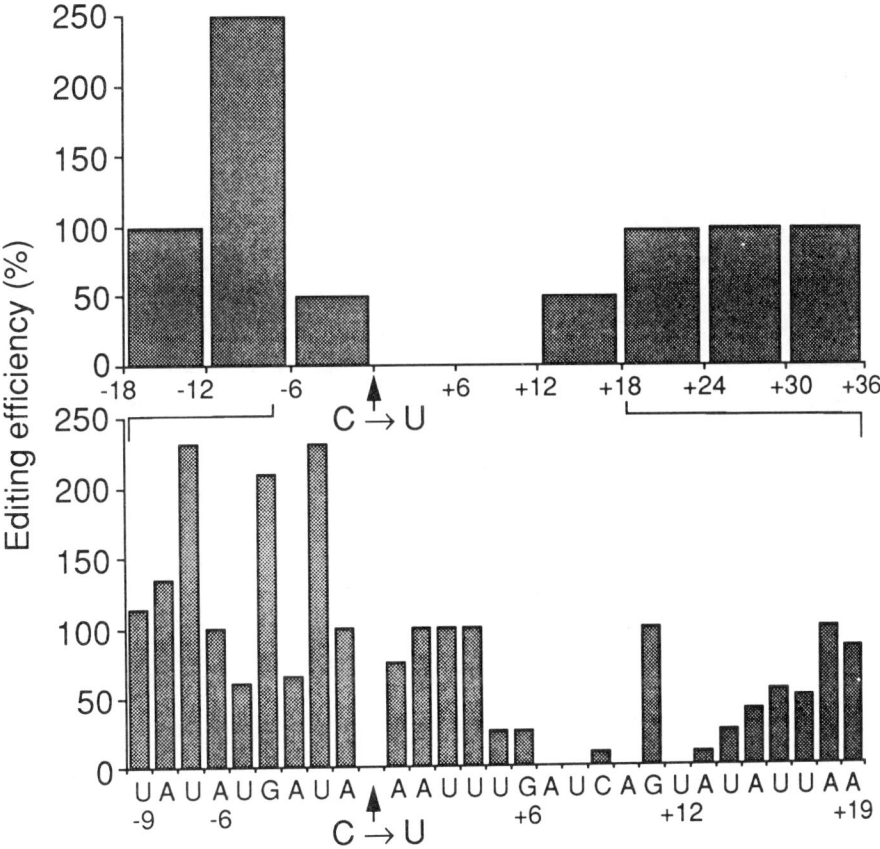

Fig. 4 — Mutagenesis of the apo B mRNA editing site. Six nucleotide scanning mutants were created to encode the complementary sequence to the native apo B mRNA (top panel). The region demonstrated by the scanning mutants to be essential for apoB mRNA editing was subjected to random and site-directed mutagenesis to create a series of point mutants (bottom panel). Nucleotides are shown on the horizontal axis and per cent editing on the vertical axis.

Table 2 —— Mutagenesis results: spacing and multiple Cs

Mutation	Sequence -3 -2 -1 0 +1 +2 +3 +4 +5	Editing at:				Simultaneous editing at:				
		-1	0	+1	+2	-1,0	0,+1	+1,+2	0,+2	0,+1,+2
	A U A C A A U U U		4+							
-A	A U A C A U U U		0							
+C	A U A C C A A U U U	2+	+			+				
A +1→C	A U A C C A U U U		4+	+			5+			
A +2→C	A U A C A C U U U		5+		2+				ND	
A +1, +2→C	A U A C C C U U U		4+	+	+		+	0	0	0

0 = 0%,
+ = <25%,
2+ = 25-50%,
3+ = 50-75%,
4+ = 75-125%,
5+ = 125-200%,
ND= data not reported.

abolished editing. When two or three cytidines are presented at the editing site, both or all are edited. However, moving the cytidines away from the target site (comparing mutations +C and A+1→C) markedly inhibits editing of both Cs. Notice that editing of an RNA containing the mutation CAA to CCA (mutation A+1→C), produces more RNA containing UUA than either UCA or CUA. Since the *in vitro* extract edits only about 2% of the mRNA in these experiments, it is unlikely that the editing enzyme binds and deaminates these RNAs twice. This implies that, once properly bound to a substrate RNA, the editing enzyme can catalyze multiple deaminations. Despite the evidence for multiple deaminations, the activity appears to be site-specific and not processive. A cytidine only two nucleotides from the editing site in the CAC mutant (mutation A+2→C) is poorly edited, and the next upstream cytidine in natural mRNA (position -11) is not edited *in vivo* or *in vitro*. The discovery of an RNA target site and the spatial restrictions between the target site and the deaminated cytidine can be accommodated by a proposal of separate RNA binding and RNA modification domains on the editing enzyme, or even separate functional subunits. The spacing between the RNA target site and the edited cytidine is fixed by the distance separating the enzyme's two binding sites. The enzyme binds to the RNA target site even after deamination of a cytidine, allowing multiple cytidines to be deaminated as a consequence of a single binding event. Structure or sequence context provided by flanking sequences can greatly promote the editing efficiency. An AU-rich context appears more favorable to editing than a CG-rich context, possibly by providing a more relaxed structure that allows access of the editing enzyme (Davies *et al.*, 1989). Backus and Smith (1991) identified the nucleotides 14-17 upstream of the editing site as affecting editing efficiency two-to four-fold. Within these nucleotides, the mammalian species differ. One of only two differences between baboon and human sequences within this 170 base region occurs at this site (-16 G versus A), and Driscoll and Casanova (1990a) report that the baboon RNA is edited three-fold more efficiently than human. Similar differences in editing efficiency have been reported between rabbit and human RNA sequences, which differ by more than 10% outside of the conserved core (Garcia *et al.*, 1992). Some of the studied point mutations or block scrambling mutations actually improve the efficiency of editing over the wild type sequence. Suboptimal efficiency may be tolerated in the natural sequence to accommodate important amino acid sequences in the apo B protein. Alternatively, suboptimal editing may allow the regulation of the level of editing to be more responsive. If editing were too efficient, it would be difficult to produce both unedited and edited transcripts simultaneously. Inefficient editing could allow regulation of the editing, by introducing a stimulatory subunit or protein modification that improves efficiency.

Recently, a second RNA editing site has been discovered in the apo B mRNA (Navaratnam *et al.*, 1991). Several sites with sequence similarity to the apo B editing site were identified elsewhere in the human apo B mRNA. These sites were tested for their *in vitro* editing by cell extracts. Most sites were unaffected by active editing extract (sites at nucleotides 5537, 6286, 6414, 6424, 6444, 6515, 6528, 6597, and 6876 as well as the site 6655 assayed by Driscoll *et al.*, 1987). One site, cytidine 6802, was edited, albeit at a lower efficiency than the primary editing site. From human intestinal cDNA libraries, 2 of 11 apo B cDNAs had a T at this site (all containing T at the primary editing site), demonstrating the use of this site *in vivo*. Fig. 3 shows that this site lies in a region of limited sequence similarity with the primary editing site, including seven of the eleven nucleotides in the target sequence identified to be the most important for editing. In addition the sequence immediately surrounding both edited cytidines is identical. Surprisingly, the spacing between the putative recognition element and the edited cytidine is expanded, which may contribute to this site's poor editing efficiency. It is unclear what effect the C to U change would have on apo B 48 mRNA's 3' untranslated region, perhaps creating an AAUAAU polyadenylation signal or creating an mRNA destabilizing element. If the second editing site is modified without editing the first, it is not clear what effect a change at codon 2198, ACA threonine to AUA isoleucine, would have on apo B100 protein, perhaps the loss of a phosphorylation site or a change in the protein's structure or lipid binding. Alternatively, the second editing site may be functionally neutral; the fortuitous recognition of a similar sequence in an mRNA targeted for editing.

In summary, the mutagenesis studies on the primary editing site and discovery of a second editing site indicate that a limited stretch of primary RNA sequence downstream of the edited nucleotides is an important determinant of RNA editing, although both apo B editing sites may be influenced by elements of sequence context, such as secondary structure, polyadenylation site proximity, AU-richness or localization on an abnormally long exon. Importantly, the mutagenesis studies and the finding of a second natural editing site dissociate RNA editing from creation of a termination codon. This form of RNA editing may be an important force for generating protein diversity, with the regulated, site-specific alteration of a protein's sequence or perhaps the generation of new initiation codons, stop codons, splice sites, polyadenylation signals, or protein binding sites on the RNA. Finally, the finding of the second edited site is a dramatic validation for the ability of the *in vitro* editing activity to reproduce its *in vivo* specificity.

6.5 THE MECHANISM OF EDITING

In vitro editing of an apo B mRNA containing $5'$-$\alpha[^{32}P]$ cytidine results in the creation of an $\alpha[^{32}P]$ uridine monophosphate residue specifically at the proper editing site (Böstrom *et al.*, 1990; Hodges *et al.*, 1991). Two important conclusions can be drawn from this simple observation. Firstly, the product of RNA editing is a simple uridine base; no complex hypermodified base is created. (A caveat is that several tRNA modifications are sequential. The C to U change reproduced *in vitro* may be the first step in a hypermodification pathway, *in vivo*.) Secondly, the $5'$-phosphate of the base is preserved. There is no known polymerase that can remove a nucleotide and add a new nucleotide in its place without replacing the $5'$-phosphate with the donor's phosphate. So, the editing reaction does not involve excision and replacement.

The process of RNA editing probably involves site-specific cytidine deamination. An alternative possibility is transglycosylation, whereby the phosphoribosyl chain is maintained, but a new base is exchanged onto the ribose. This reaction mechanism is used to introduce hypoxanthine (Elliott and Trewyn, 1984) and queosine (Okada and Nishimura, 1979; Okada *et al.*, 1979) into pre-formed tRNAs. No donor base can be identified though, and there is no component to the editing activity lost by dialysis or filtration. Finally, it should be kept in mind that the *in vivo* activity could be transamination, capturing the amine to transfer it to another molecule or perhaps catalyzing the forward and reverse reactions, interconverting U and C residues. However, this also is unlikely, since the enzyme does not require a pyridoxal cofactor as do many transaminases and appears capable of catalyzing multiple deaminations while bound to the RNA (see above).

Incorporation of modified nucleotides into the RNA substrate can have profound consequences on *in vitro* editing (Hodges and Scott, unpublished). Incorporation of either CTP or UTP substituted at the $5'$ position with a halogen (bromine or iodine) or a mercurisulfhydryl group creates an RNA which is unable to be edited. The effect is not strictly steric, as incorporation of 5-allylamine uridine or its biotinylated derivative bio-11-uridine has no effect on editing. The inhibitory modifications may have specific effects on the editing enzyme, but this cannot be determined in such a crude system. The editing enzyme may also contact the RNA via phosphoryl and ribosyl groups distant from the edited nucleotide: adenosyl thiophosphorylate linkages inhibit editing, even when not immediately adjacent to the edited nucleotide, and 2'-deoxycytidine incorporation inhibits editing too. Further, editing of unmodified RNA is inhibited by vanadyl ribonucleosides, but not free vanadyl sulfate, implying a contact between the editing enzyme and the substrate's phosphates. For all these modifications, it is not known which specific nucleotide modifications are inhibitory, if the modification alters specific groups that interact with the enzyme, or if the effect is on the RNA's conformation.

6.6 CHARACTERIZATION OF THE RNA EDITING ACTIVITY

The interaction of apo B mRNA and cellular proteins has been investigated by non-denaturing gel electrophoresis of the RNA-protein complexes (RNA gel mobility shift; Lau *et al.*, 1990, Shah *et al.*, unpublished), by glycerol gradient sedimentation of the reaction components (Smith *et al.*, 1991; Greeve *et al.*, 1991), and by RNA-protein crosslinking by UV light (Lau *et al.*, 1990; Shah *et al.*, unpublished). The mobility shift assay shows that proteins in rat liver or hepatoma extracts bind specifically to apo B RNAs. Smith *et al.* (1991) have demonstrated by gradient sedimentation that the apo B RNA substrate progressively assembles over 3 hours with the rat liver proteins into 11S and then 27S complexes, and it has been proposed that the latter is the active "editosome" - after the term spliceosome. Greeve *et al.* (1991) have also demonstrated the free editing activity from rat enterocytes sediments at 11S, but this extract requires no lag period prior to catalysis, and they were unable to show assembly into a higher order complex on the RNA substrate. UV light crosslinks one 40 kDa protein to apo B mRNA, with specificity for the editing site (Lau *et al.*, 1990). Shah *et al.* (unpublished) have crosslinked several proteins to the RNA (85 kDa, 64 kDa, 50 kDa, and 35 kDa). At least two of these proteins bind specifically to the apo B RNA editing site and are both present in liver and intestine in related forms. However, there is as yet no direct connection of the deaminase activity with either the crosslinking proteins or the RNA-protein assemblies, and it has not yet been demonstrated that any of these RNA binding proteins bind specifically to the editing target site.

An initial biochemical characterization of the RNA editing activity has been carried out (Driscoll and Casanova, 1990b; Greeve *et al.*, 1991; Garcia *et al.*, 1992). Editing *in vitro* is optimal in 50-150 mM NaCl or KCl, at around pH 8, at 30°C. It is detected after as little as five minutes, gives a linear accumulation of product for 3-5 hours, and can continue at a reduced rate for at least 48 hours. It shows enzymatic kinetics, with the apparent k_m for the RNA editing site of 0.4 nM (Garcia *et al.*, 1992) to 2 nM (Greeve *et al.* 1991). While the differing k_m values are likely to reflect different affinity for rabbit versus human RNA, these affinities are 1000-fold higher than the affinity of the classical cytosine or cytidine deaminase for their substrate (Garcia *et al.*, 1992). This is likely to result from the larger number of contacts an enzyme can make with an RNA target site than with a single nucleotide, and could allow the enzyme to locate efficiently the RNAs to be edited during their maturation in the nucleus.

The editing extracts are active after dialysis, ammonium sulfate precipitation, and gel filtration, indicating no requirement for low molecular weight cofactors or a second substrate. Activity is not stimulated by nucleotides or an ATP regeneration system. It does not require divalent cations, but is stimulated by some (2 mM calcium or magnesium) and inhibited by others (2 mM iron, nickel, cobalt, zinc, copper or cadmium) and many preparations can be stimulated by chelating the endogenous inhibitory ions. The possibility that the enzyme contains a buried zinc atom, as does adenosine deaminase, is discussed below. Activity can be stimulated modestly by heparin (25 ng/ul), glycerol (40%), ethylene glycol (20%), or polyethylene glycol (15%). The activity is not inhibited by tetrahydrouracil, a transition-state analog for cytosine deamination (Driscoll and Casanova, 1990a), implying the active deamination site may be hidden until properly bound to RNA. It is inhibited by >1 mM vanadyl ribonucleoside complexes, ribonucleotide analogues which are traditionally used to inhibit ATPases and phosphodiester transferases, but not by free vanadyl sulfate (Smith *et al.*, 1991). Since the enzyme does not require ATP and does not cleave the RNA's phosphodiester backbone, this implies that the vanadyl ribonucleosides inhibit binding to the RNA target.

The editing activity of crude S100 cytosolic extracts is destroyed by protease treatment or heating to 40°C for 10 minutes, and by sulfhydryl modification with N-ethyl maleimide or p-mercuribenzoate, imidazole modification by diethylpyrocarbonate, or guanidino modification by phenylglyoxal. The activity is not destroyed by ribonuclease, deoxyribonuclease or non-specific nuclease (although nuclease digestion products from crude extracts can inhibit the reaction (Greeve *et al.*, 1991). The buoyant density of partially purified editing enzyme was 1.3 g/ml, that is of pure protein. Therefore, the apolipoprotein B mRNA editing activity consists of protein with no RNA component.

Progress in purification of the enzyme has been slowed by the limited tissue source (intestinal enterocytes) and the apparent low level of the enzyme in the tissue. Partial purification has involved ammonium sulfate precipitation (at between 20-40% saturation), adsorption to DEAE ion exchange resin (eluted at 150 mM KCl), gel filtration (although estimations of its size have ranged from 90 kDa, to 125 kDa, to 150-200 kDa), and glycerol gradients (sedimenting at 11S). The activity can also be recovered from cesium chloride density gradients (at 1.3 gm/ml).

In summary, the apo B editing enzyme is a protein(s), with no detectable RNA component. The active site may contain important cysteine, histidine and arginine residues, and binding of the RNA phosphodiester backbone may be inhibited by vanadyl ribonucleoside complexes. It binds to the RNA target with high affinity and is immediately competent for editing without a lag period for assembly. There is no requirement for a donor uridine nucleotide. It does not seem to contain a pyridoxal cofactor as do many transaminases, and is not inhibited by tetrahydrouridine as is cytidine deaminase. On glycerol gradients the editing enzyme sediments at 11S (240 kD) or larger, but has a lower apparent molecular weight on gel filtration (about 125 kD). The relationship of the catalytic component of the enzyme to other cellular proteins that bind the RNA substrate specifically has yet to be clarified. However, in its most purified form it appears to have a simple composition, and bears no similarity to the spliceosomal complex.

6.7 THE REGULATION OF APO B mRNA EDITING

Apo B mRNA editing is tissue specific, and hormonally and developmentally regulated. The apo B gene is expressed primarily in liver parenchymal hepatocytes and intestinal epithelial enterocytes, and to a lesser extent in kidney proximal tubules, placenta, and embryonic visceral yolk sac. Accordingly, the principal tissue culture cell models have been rat hepatoma cell lines (McArdle RH7777 and Fao) which express endogenous apo B and edit about 20%, and the CaCo-2 human colonic adenocarcinoma line which expresses apo B but does not edit until induced to differentiate into a small intestinal enterocyte-like form (Jiao et al., 1990). Among different species, the extent of apo B mRNA editing in the same tissue can be very different. Human liver mRNA is largely unedited, while in rodents the liver mRNA is edited to a substantial percentage. This difference has been cited as one of the explanations for elevations in circulating LDL cholesterol seen in humans relative to rodents.

With more sensitive detection afforded by PCR amplification, apo B transcripts were detected in many mouse tissues (Davies et al., 1989). It is not clear whether this low level expression is physiologically important or represents aberrant transcription, but it does give an opportunity to measure editing activity in a wide range of tissues. The proportion of apo B transcripts that are edited can vary considerably among different tissues (intestine 98%, liver 57%, kidney 52%, brain 70%, spleen 70%, testis 44%), yet most tissues contain editing activity. This result raises the question of how to view the regulation of editing: is editing induced in the intestine or repressed in the liver? The trace of apo B mRNA detected in various tissues is edited to very different extents. This suggests that the expression of editing activity in most tissues is unrelated to apo B expression. While apo B mRNA editing is the only assay we have for this unusual activity, it seems likely that this RNA editing acts on other mRNAs, and its level is regulated in tissue specific patterns. This has been most convincingly demonstrated by transfection of an apo B minigene into cultured cells (Böstrom et al., 1990). Although a wide range of cell lines show little or no editing activity, several osteosarcoma cell lines (most notably G-292 cells) and one epidermoid cell line (A431), which do not express endogenous apo B, can edit the transgene RNA. It is unknown what endogenous RNA targets there are for these editing activities.

Editing activity is induced in the development of rat and human intestinal enterocytes and rat hepatocytes. The developmental profile of rat apo B mRNA editing is shown in Fig. 5. Both these tissues express only unedited apo B mRNA early in development. In rats (Baum et al., 1990), pigs (Teng et al., 1990a) and humans (Teng et al., 1990b), the intestinal mRNA is edited to a progressively greater extent around birth, reaching levels of nearly complete editing in adults. In the liver of rodents, editing is also induced, but on an independent timecourse and reaching a lower final plateau. Thus, these changes most probably represent true developmental patterns and not mere reflections of changing hormonal or nutritional status. This view is supported by the ability to induce apo B editing upon *in vitro* differentiation of the CaCo-2 cell line. This tissue culture cell model will be important to study the regulation of the RNA

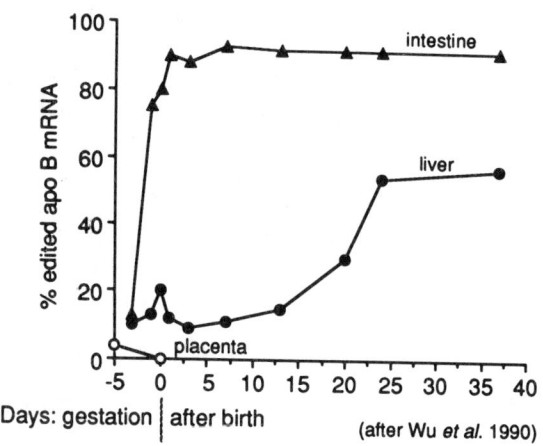

Fig. 5 — The developmental profile for editing activity in rat intestine, liver and placenta.

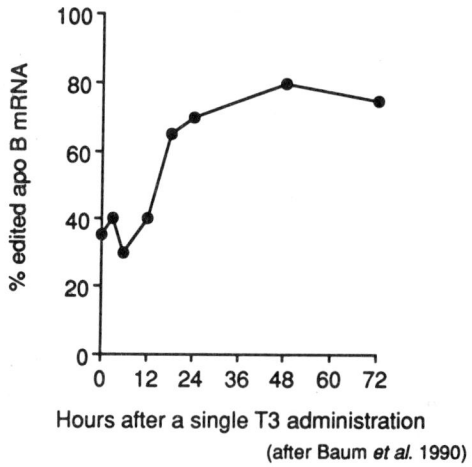

Fig. 6 — Induction of liver apo B mRNA editing in hypothyroid rats by T_3 administration.

editing activity in development.

The editing of apo B mRNA is under hormonal and nutritional regulation as well. In rats, fasting induces a modest decrease in hepatic apo B mRNA editing; refeeding with a carbohydrate-rich diet induces a dramatic increase in editing to nearly completely edited message. Thyroid hormone is required for editing in the rat liver, but not in the intestine (Patterson *et al.*, 1991; Baum *et al.*, 1990). Removing the thyroid glands decreases the editing of apo B mRNA in the liver, and subsequent administration of the thyroid hormone T_3 induces editing (Fig. 6). Both of these effects are coordinated with induction of lipogenesis, and the effect of the hormones on editing activity may be indirect. However, the effects of thyroid hormone are mediated directly by nuclear thyroid hormone receptors (Patterson *et al.*, 1991) and are modified secondarily by growth hormone (Sjöberg *et al.*, 1992). Most importantly for further study, the effects of hormonal modulation on the levels of editing activity are maintained in soluble editing extracts made from treated rats (Harris and Smith, 1992), and can be reproduced by hormone treatment of isolated hepatocytes (Sjöberg *et al.*, 1992). So, it can be determined whether the regulation of editing activity is brought about by increasing the quantity of the editing enzyme or by stimulating the activity of existing enzyme, and the signal transduction pathway that regulates this hierarchy can be defined. As an indication of the power of these systems for studying the regulation of the editing activity, it has been reported that the HepG2 human hepatoma cell line and the colonic CaCo-2 line, which do not normally express editing activity, can be induced to edit by transfection with a thyroid hormone receptor gene and a growth hormone gene (Patterson *et al.*, 1991). These hormones turn on the editing activity in cells which are competent to respond.

6.8 A MORE GENERAL ROLE FOR RNA EDITING IN THE REGULATION OF GENE EXPRESSION?

The apo B mRNA editing activity is induced in development and by hormonal and dietary responses. These changes can be rationalized as responding to increased demand for lipid secretion, and the induction of apo B editing is coordinated with induction of lipogenic enzymes. If apo B mRNA is not the sole target for this RNA deaminase, it is likely that the mRNAs encoding other proteins of this lipogenic response are edited to encode high response isozymes. What other proteins are regulated and how?

The developmental and hormonal response of the liver and the intestine are quite distinct. This implies that the RNA deaminase is independently regulated in different tissues. While lipogenesis is regulated by editing in the liver, an entirely different pathway may be regulated by editing in the brain, spleen, or testes.

If the mRNA editing that creates apo B48 is a general RNA processing pathway, we can ask what the consequences of this editing could be and how editing could coordinate changes in multiple targets. A discrete C to U deamination could create a new initiation codon (ACG to AUG), create termination codons (CAA, CAG, CGA to UAA, UAG, UGA), or create missense

Table 3 — RNA editing and gene expression

First position changes

Leu→Phe	Leu→Leu	Pro→Ser	His→Tyr
CUY→UUY	CUR→UUR	CCX→UCX	CAY→UAY
Gln→TER	Arg→Cys	Arg→TER	Arg→Trp
CAR→UAR	CGY→UGY	CGA→UGA	CGG→UGG

Second position changes

Ser→Phe	Ser→Leu	Pro→Leu	Thr→Ile
UCY→UUY	UCR→UUR	CCX→CUX	ACY→AUY
Thr→Ile	Thr→Met	Ala→Val	
ACA→AUA	ACG→AUG	GCX→GUX	

Third position changes

All changes are silent

---

alterations (see Table 3).

The resultant amino substitutions vary in their severity: potentially drastic changes (such as losing proline kinks, removing charged amino acids, or hydrophobic for hydrophilic changes); changes that could have profound regulatory potential (such as removing or creating sites for phosphorylation, glycosylation, fatty acid attachment, propeptide cleavage, or disulfide bonding); or changes that are neutral at the protein level. It should be kept in mind that "silent" codon changes can still have profound changes, for instance by limiting translation rates by requiring a rare tRNA.

The effect of RNA editing could be manifest on the RNA itself, for instance to create an RNA sequence which affects the processing, translation or stability of the RNA. Most importantly, it must be kept in mind that apo B RNA editing can be a nuclear event and can precede splicing and polyadenylation. It is conceivable, for instance, that an intronic editing target sequence directs the conversion of an inefficient 5' splice site beginning with GC to the consensus GU, enhancing the production of mature mRNA. This editing event would escape detection, since the edited base is spliced out and degraded in the nucleus. Similarly, the GU-rich or U-rich elements downstream of polyadenylation sites could be enhanced by RNA editing, and the evidence of the editing would remain hidden in the nucleus. More obviously, editing can create target sequences in the RNA, to create an AAUAAA polyadenylation signal or an

AU-rich destabilizing element, to enhance the context of a poorly used initiation codon, to remove or create secondary structure, or to create or destroy protein binding sites such as for the iron response element. This last scenario imposes a combinatorial regulation on the target RNA: inducing RNA editing shifts the mRNA under additional regulatory hierarchy. Finally, the RNA editing target site UGAUCAGUAUA is more than five times as efficient as UGAUCAGCAUA. Editing could create another editing site; tandem editing sites could change the regulatory response from a simple linear characteristic to a logarithmic response.

6.9 THE ORIGINS OF RNA EDITING BY DEAMINATION

All of the non-insertional RNA editing activities so far discovered could be (or have been demonstrated to be) site-specific RNA base deaminases or transaminases. The apo B mRNA editing enzyme is apparently a cytidine deaminase. Both the mitochondria and the chloroplasts of higher plants convert single or multiple cytidine residues to be read as uridines, although the reverse reaction of C to U conversion is also seen in mitochondria (see Chapter 7). Selenocysteine inserting tRNAs undergo a pattern of C to U and U to C conversions (Diamond *et al.*, 1990). The unwinding of double stranded RNA (dsRNA) is accomplished by a dsRNA specific adenosine deaminase, which converts adenosyl residues to inosines (see Chapter 8). The RNA editing that changes a specific CAG glutamine codon to an arginine codon in the mammalian neuronal glutamate-gated ion channel mRNA (Sommer *et al.*, 1991) may also be a deaminase, creating a CIG codon.

  Why should so many different RNA editors operate by site-specific deamination? The answer is that these systems are either divergent (derived from a common ancestral editor) or convergent (utilizing a similar solution to similar problems). The first proposal, that these RNA deaminases share a common heritage, is not easily evaluated. Since similar double stranded RNA deaminases exist in nematodes, insects and vertebrates, they must have existed since the radiation of metazoan life, at least 600 million years. The origins of the apo B mRNA deaminase are less easily placed. Apolipoprotein B probably evolved within the vertebrate lineage from a vitelogenin-like lipid carrier (Baker, 1988). Editing of the apo B mRNA was probably acquired within the mammalian line, as neither chick nor frog apo B mRNA is edited (N. Davidson, personal communication). While all these species have reasonably homologous genes, the mammalian genes diverge from the other vertebrates in the region of the editing site. This leads to the conclusion that either the apo B mRNA deaminase evolved recently in mammals or that it pre-existed and mammalian evolution introduced a new editing site into the apo B gene. We favor the latter hypothesis, since the editing activity appears to be independent of apo B expression in most tissues. Since we cannot use apo B mRNA editing to date the origins of this deaminase, we must await the discovery of another, more ubiquitous target for the deaminase. Still, it is illustrative to imagine apo B editing as a paradigm of evolution shifting an existing RNA editing activity to a new gene target, creating a new gene product from an existing gene and introducing another gene into a common regulatory pathway.

To support the alternative proposal, that the different RNA editing systems are convergent, it must be argued that 1) the solution of deamination is a common one, 2) the facilities to deaminate nucleotides are present in many organisms, and 3) the evolution of a deaminating system might be more advantageous than other modification pathways. These conditions are easily met: nucleotide deaminases are essential genes carried by most organisms; generating specific editing systems may be as easy as duplicating a deaminase gene and combining it with the gene for an RNA binding protein; and editing systems are free from the restraints of energy consumption and cofactor availability.

Many pathways of site-specific RNA modification exist, to introduce modified or hypermodified bases into tRNAs, rRNAs, snRNAs and even to modify mRNAs by methylation. While these modifications can affect the stability of the RNA or modulate the nucleotide's conformation so as to allow more stringent or more flexible base pairing, as a rule these modifications do not alter the essential identity of a nucleotide. Modified cytidine residues still base pair most effectively with guanosine. Only one tRNA hypermodification has been demonstrated to alter the identity of a nucleotide. In the anticodon of the minor AUA-recognizing tRNA^{Ile}, the attachment of a lysine amino acid to a cytidine forms lysidine, which base pairs with adenosine exclusively (Muramatsu et al., 1988). This pathway of tRNA modification is conserved among archebacteria, eubacteria, mitochondria, and chloroplasts, and represents an interesting corollary to the mRNA editing systems.

It is surprising that no other pathways of induced nucleotide mutation have been identified. In contrast, the mutagenic properties of numerous chemicals demonstrate that many chemical adducts are base-specific and can alter the coding capacity of the nucleotide. RNA editing by deamination may have been selected because it offers a number of simplifications over base modification: no genotoxic adduct needs to be synthesized, the enzyme needs no second substrate except water, the reaction is energetically favorable requiring no energy input, and a progenitor for this editing enzyme exists in all organisms as the enzymes of nucleotide metabolism.

The conversion of cytosine and cytidine nucleoside derivatives to their corresponding uracil derivatives is accomplished by a number of enzymes specific for different roles in the biosynthesis and degradation of nucleic acids. This is because all pyrimidines are synthesized via a common pathway and degraded by a common pathway. Cytidine triphosphate is made from UTP, by transamination with glutamine in the ATP-dependent reaction of CTP synthetase. Degradation of the cytosine base, cytidine nucleoside and nucleotides is catalyzed by deamination to uracil or uridine derivatives by cytosine deaminase, cytidine deaminase, dCMP deaminase, or dCTP deaminase. These enzymes also function to interchange nucleotide pools and in the salvage pathway to incorporate preformed bases. Similarly, all purine biosynthesis derives from the initial synthesis of inosine triphosphate. Adenine, adenosine and adenosine nucleotides (but not guanine derivatives) are degraded through deamination to inosine. The nucleotide, nucleoside and base deaminases vary considerably in their subunit molecular weight (20 to 93 kDa) and multimer composition (from

Table 4 —— Comparison of nucleotide deaminase subunit structure

Deaminase	Species	Native MW	Subunit MW
Cytosine	Yeast	34 kDa	34 kDa
Cytidine	Yeast	57 kDa	NK
	E. coli	73 kDa	NK
	Mouse	74 kDa	NK
dCMP	Yeast	NK	36 kDa
	T2/T4 phage	124 kDa	21 kDa
dCTP	Salmonella	82 kDa	NK
Adenosine	E. coli	36 kDa	36 kDa
	Mammals	36 kDa	36 kDa
		>200 kDa	36 kDa + complexing protein
AMP	Yeast	NK	93 kDa
	Rabbit	320 kDa	80 kDa
Adenosine (non-specific)	Aspergillus	215 kDa	30 kDa
Guanine	Rabbit	170 kDa	NK
Guanosine	Pseudomonas	100-200 kDa	NK

NK= not known.

dimers to hexamers) (see Table 4). Since deaminase enzymatic activity can be carried by a relatively small protein domain, an RNA editing activity may have a related subunit or domain joined to a targeting protein responsible for RNA recognition. The deaminases proposed to edit apo B mRNA, dsRNA, glutamate receptor channel mRNA, tRNA[Ser]Sec and plant mitochondrial and chloroplast RNA may have a similar structure, or even share a catalytic subunit.

Mechanistic similarities have been demonstrated between cytidine or dCMP deaminase and adenosine or AMP deaminase (Frick et al., 1989). Both classes of enzymes catalyze energetically favorable hydrolytic deaminations with remarkable rate enhancement. Both have high affinity for hydrated substrate analogues, suggesting a mechanism of water addition across the carbon-nitrogen double bond (cytidine C^4-N^3) followed by or coupled with ammonia elimination (see Fig. 7). For both, histidine and cysteine residues are required for activity, as they are for the apo B editing activity as well as the dsRNA adenosine deaminase. Solving the X-ray crystallographic structure of adenosine deaminase uncovered some surprises (Wilson et al., 1991). Most importantly, a zinc atom is held in the active site. The zinc cofactor was unexpected, as it is not removed from the enzyme by dialysis, filtration, enzyme purification or EDTA treatment. Binding of zinc to the attacking water molecule provides the remarkable affinity of the enzyme for the hydrated intermediate. A hidden zinc ion is likely to be found at the active site of cytidine deaminase, which has a similarly high affinity for its hydrated intermediate. A tightly bound metal ion may explain why both apo B mRNA cytidine deaminase and the dsRNA adenosine deaminase are inhibited by certain metals and are stimulated in crude extracts by high concentrations of EDTA.

The active site residues of adenosine deaminase are found in regions of limited primary sequence homology with AMP deaminase (Chang *et al.*, 1991) which on further inspection resembles homologous regions of dCMP deaminase. The regions of highest homology are the histidines which coordinate the zinc ion, the active site acidic residues, and the hydrophobic "lid" over the active site. It can be proposed that these deaminases share a common reaction mechanism, reflected in divergent or convergent similarities in their amino acid sequence, and that the RNA editing deaminases may share these similarities.

The two best characterized site-specific RNA deaminases, the apo B mRNA deaminase and the dsRNA deaminase, share a number of interesting (or perhaps coincidental) similarities (see also Chapter 8). Both recognize an RNA structure and perform a hydrolytic base deamination with no energy or cofactor requirement. However, neither activity is inhibited by the classic nucleotide deaminase transition state analogue inhibitors (coformycin for adenosine deaminase and tetrahydrouracil for cytidine deaminase). Access to the active site may be hidden in both RNA deaminases until RNA is properly bound. Both RNA deaminases can deaminate multiple bases on the target RNA; although the dsRNA deaminase is thought to extensively modify its target, recent experiments show that the enzyme can act as a site-specific deaminase on RNAs with short imperfect duplexes (Sharmeen *et al.*, 1991). Both are proteinaceous, although both may be bound to RNA in crude extracts: the dsRNA deaminase co-purifies with inosine containing RNA and the apo B mRNA deaminase in crude preparations is inhibited by RNase and has an isoelectric point that is strongly acidic. Both are inhibited by sulfhydryl modification and, more interestingly, both are inhibited by vanadyl ribonucleosides although neither are ATPases or phosphatases. Both are stimulated by high levels of metal chelators but only in crude preparations; free zinc inhibits the dsRNA deaminase at micromolar concentrations and the apo B mRNA deaminase at millimolar concentrations. Both bind to DEAE ion exchange columns and are eluted by about 200 mM KCl. The dsRNA deaminase sizes as 210 kDa in gel filtration and 9-10S by sedimentation; the apo B mRNA deaminase sizes as 125 kDa but sediments as 11S. From preliminary results, both activities contain only one or a few proteins which are cross-linked to the substrate RNA by UV light. For both activities the protein-RNA binding appears to be much more rapid than the subsequent deamination. Given the similarities of these two enzymes, it is not unreasonable to propose that all RNA editing deaminases share mechanistic similarities, or descend from a common progenitor, or even share common protein subunits.

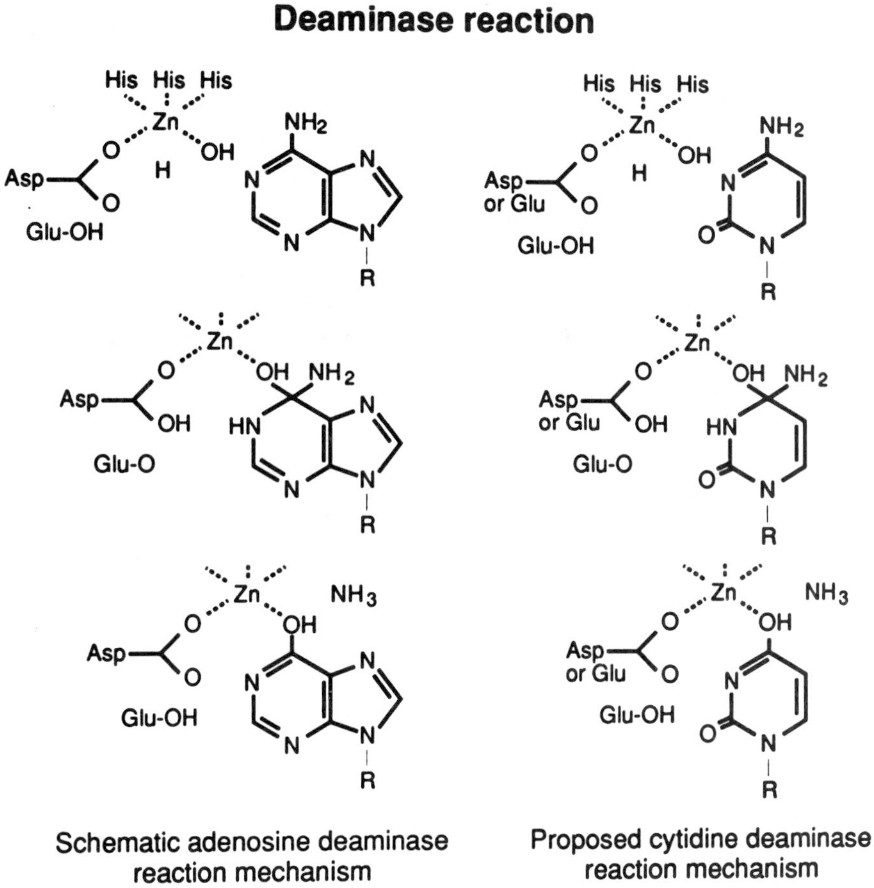

Deaminase reaction

Schematic adenosine deaminase
reaction mechanism

Proposed cytidine deaminase
reaction mechanism

Fig. 7 — Schematic representation of the cytidine deaminase reaction based on the known homologies between soluble cytidine deaminase and adenosine deaminase, in which the presence of zinc in the active site has been demonstrated.

REFERENCES

Aburatani H., Matsumoto A., Ishikawa T., Takaku F. & Itakura H. (1989) Single base substitution between human intestinal and hepatic apolipoprotein B mRNA detected by ribonuclease cleavage analysis. *J Biochem Tokyo* **105** 911-915.

Backus J. W. & Smith H. C. (1991) Apolipoprotein B mRNA sequences 3´ of the editing site are necessary and sufficient for editing and editosome assembly. *Nucl Acids Res* **19** 6781-6786.

Baker M. E. (1988) Is vitellogenin an ancestor of apolipoprotein B-100, of human low-density lipoprotein and human lipoprotein lipase? *Biochem J* **255** 1057-1060.

Baum C. L., Teng B.-B. & Davidson N. O. (1990) Apolipoprotein B messenger RNA editing in the rat liver. *J Biol Chem* **265** 19263-19270.

Böstrom K., Lauer S. J., Poksay K. S., Garcia Z., Taylor J. M. & Innerarity T. L. (1989) Apolipoprotein B48 RNA editing in chimeric apolipoprotein EB mRNA. *J Biol Chem* **264** 15701-15708.

Böstrom K., Poksay K. S., Johnson D. F., Lusis A. J. & Innerarity T. L. (1990) Apolipoprotein B mRNA editing. Direct determination of the edited base and occurrence in non-apolipoprotein B-producing cell lines. *J Biol Chem* **265** 22446-22452.

Chang Z., Nygaard P., Chinault A. C. & Kellems R. E. (1991) Deduced amino acid sequence of *Escherichia coli* adenosine deaminase reveals evolutionarily conserved amino acid residues: implications of catalytic function. *Biochemistry* **30** 2273-2280.

Chen S.-H., Habib G., Yang C. Y., Gu Z. W., Lee B. R., Weng S. A., Silberman S. R., Cai S. J., Deslypere J. P., Rosseneu M., Gotto A. M. Jr, Li W. H. & Chan L. (1987) Apolipoprotein B-48 is the product of a messenger RNA with an organ-specific in-frame stop codon. *Science* **238** 363-366.

Chen S.-H., Li X., Liao W.S., Wu J.H. & Chan L. (1990) RNA editing of apolipoprotein B messenger RNA - sequence specificity determined by *in vitro* coupled transcription editing. *J Biol Chem* **265** 6811-6816.

Davies M. S., Wallis S. C., Driscoll D. M., Wynne J. K., Williams G. W., Powell L. M. & Scott J. (1989) Sequence requirements for apolipoprotein B RNA editing in transfected rat hepatoma cells. *J Biol Chem* **264** 13395-13398.

Davies M. S., Cooper P. J., Wynne J. K., Driscoll D. M. & Scott J. (1990) Expression and editing of apolipoprotein-B mRNA in a wide variety of mouse tissues. *Arteriosclerosis* **10** 778a.

Diamond A. M., Montero-Puerner Y., Lee B. J. & Hatfield D. (1990) Selenocysteine inserting tRNAs are likely generated by tRNA editing. *Nucl Acids Res* **18** 6727.

Driscoll D. M., Wynne J. K., Wallis S. C. & Scott J. (1989) An in vitro system for the editing of apolipoprotein B mRNA. *Cell* **58** 519-525.

Driscoll D. M. & Casanova E. (1990a) *In vitro* editing of apolipoprotein B mRNA. *Arteriosclerosis* **10** 752a.

Driscoll D. M. & Casanova E. (1990b) Characterization of the apolipoprotein B mRNA editing activity in enterocyte extracts. *J Biol Chem* **265** 21401-21402.

Elliott M. S. & Trewyn R. W. (1984) Inosine biosynthesis in transfer RNA by an enzymatic insertion of hypoxanthine. *J Biol Chem* **259** 2407-2410.

Frick L., Yang C., Marquez V. E. & Wolfenden R. (1989) Binding of pyrimidin-2 one ribonucleoside by cytidine deaminase as the transition-state analogue 3,4-dihydrouridine and the contribution of the 4-hydroxyl group to its binding affinity. *Biochemistry* **28** 9423-9430.

Garcia Z. C., Poksay K. S., Bostrom K., Johnson D. F., Balestra M, E., Shechter I. & Innerarity T. L. (1992) Characterization of apolipoprotein B mRNA editing from rabbit intestine. *Atheroscler Thrombosis* **12** 172-179.

Greeve J., Navaratnam N. & Scott J. (1991) Characterization of the apolipoprotein B mRNA editing enzyme: no similarity to the proposed mechanism of RNA editing in kinetoplastid protozoa. *Nucl Acids Res* **13** 3569-3576.

Harris S. G. & Smith H. C. (1992) *In vitro* apolipoprotein B mRNA editing activity can be modulated by fasting and refeeding rats with a high carbohydrate diet. *Biochem Biophys Res Comm* **183** 899-903.

Hawkins J. D. (1988) A survey on intron and exon lengths. *Nucl Acids Res* **16** 9893-9908.

Higuchi K., Hospattankar A. V., Law S. W., Meglin N., Cortright J. & Brewer H. B. Jr (1988) Human apolipoprotein B (apoB) mRNA: identification of two distinct apoB mRNAs, an mRNA with the apoB-100 sequence and an apoB mRNA containing a premature in-frame translational stop codon, in both liver and intestine. *Proc Natl Acad Sci USA* **85** 1772-1776.

Hodges P. E., Navaratnam N., Greeve J. C. & Scott J. (1991) Site-specific creation of uridine from cytidine in apolipoprotein B mRNA. *Nucl Acids Res* **19** 1197-1201.

Hospattanker A. V., Higuchi K., Law S. W., Meglin N. M. & Brewer H. B. Jr (1987) Identification of a novel in-frame translational stop codon in human intestine apo B mRNA. *Biochem Biophys Res Comm* **148** 279-286.

Jenh C.-H., Deng T., Li D., DeWille J. & Johnson L. F. (1986) Mouse thymidylate synthase gene. *Proc Natl Acad Sci USA* **83** 8482-8486.

Lau P. P., Chen S.-H., Wang J. C. & Chan L. (1990) A 40 kilodalton rat liver nuclear protein binds specifically to apolipoprotein B mRNA around the RNA editing site. *Nucl Acids Res* **18** 5817-5821.

Lau P. P., Xiong W., Zhu H.-J., Chen S.-H. & Chan L. (1991) Apolipoprotein B mRNA editing is an intranuclear event that occurs posttranscriptionally coincident with splicing and polyadenylation. *J Biol Chem* **266** 20550-20554.

Muramatsu T., Yokoyama S., Horie N., Matsuda A., Ueda T., Yamaizumi Z., Kuchino Y., Nishimura S. & Miyazawa T. (1988) A novel lysine-substituted nucleoside in the first position of the anticodon of minor isoleucine tRNA from *Escherichia coli. J Biol Chem* **263** 9261-9267.

Naora H. & Deacon N. J. (1982) Relationship between the total size of exons and introns in protein-coding genes of higher eukaryotes. *Proc Natl Acad*

Sci USA **79** 6196-6200.

Navaratnam N., Patel D., Shah R. R., Greeve J. C., Powell L. M., Knott T. J. & Scott J. (1991) An additional editing site is present in apolipoprotein B mRNA. *Nucl Acids Res* **19** 1741-1744.

Ojala D., Montoya J. & Attardi G. (1981) tRNA punctuation model of RNA processing in human mitochondria. *Nature* **290** 470-474.

Okada N. & Nishimura S. (1979) Isolation and characterization of a guanine insertion enzyme, a specific tRNA transglycosylase, from *Escherichia coli. J Biol Chem* **254** 3061-3066.

Okada N., Noguchi S., Kasai H., Shindo-Okada N., Ohgi T., Goto T. & Nishimura S. (1979) Novel mechanism of post-transcriptional modification of tRNA. *J Biol Chem* **254** 3067-3073.

Patterson A. P., Eggerman T. L., Demosky S. J. & Brewer H. B. Jr (1991) T_3 nuclear receptor and growth hormone modulate apolipoprotein B editing. *Arterioscler Thromb* **11** 1403a.

Powell L. M., Wallis S. C., Pease R. J., Edwards Y. H., Knott T. J. & Scott J. (1987) A novel form of tissue-specific RNA processing produces apolipoprotein-B48 in intestine. *Cell* **50** 831-840.

Robberson B. L., Cote G. J. & Berget S. M. (1990) Exon definition may facilitate splice site selection in RNAs with multiple exons. *Mol Cell Biol* **10** 84-94.

Shah R. R., Knott T. J., Le Gros J. E., Navaratnam N., Greeve J. C. & Scott J. (1991) Sequence requirements for the editing of apolipoprotein B mRNA. *J Biol Chem* **266** 16301-16304.

Sharmeen L., Bass B., Sonenberg N., Weintraub H. & Groudine M. (1991) Tat-dependent adenosine-to-inosine modification of wild-type transactivation response RNA. *Proc Natl Acad Sci USA* **88** 8096-8100.

Sjöberg A., Oscarsson J., Böstrom K., Innerarity T. L., Edén S. & Olofsson S.-O. (1992) Effects of growth hormone on apolipoprotein B (apo B) messenger ribonucleic acid editing and apo B48 and apo B100 synthesis and secretion in the rat liver. *Endocrinology* **130** 3356-3364.

Smith H. C., Kuo S. R., Backus J. W., Harris S. G., Sparks C. E. & Sparks J. D. (1991) *In vitro* apolipoprotein B mRNA editing: identification of a 27S editing complex. *Proc Natl Acad Sci USA* **88** 1489-1493.

Sommer B., Köhler M., Sprengel R. & Seeburg P. H. (1991) RNA editing in brain controls a determinant of ion flow in glutamate-gated channels. *Cell* **67** 11-19.

Teng B., Black D. D. & Davidson N. O. (1990a) Apolipoprotein B messenger RNA editing is developmentally regulated in pig small intestine: nucleotide comparison of apolipoprotein B editing regions in five species. *Biochem Biophys Res Comm* **173** 74-80.

Teng B., Verp M. & Davidson N. (1990b) Apolipoprotein B (apo B) mRNA editing is developmentally regulated and widely expressed in human tissues. *Arteriosclerosis* **10** 752a.

Wilson D. K., Rudolph F. B. & Quiocho F. A. (1991) Atomic structure of adenosine deaminase complexed with a transition-state analog: understanding catalysis and immunodeficiency mutations. *Science* **252** 1278-1284.

Wu J. H., Semenkovich C. F., Chen S.-H., Li W.-H. & Chan L. (1990) Apolipoprotein B messenger RNA editing - validation of a sensitive assay and developmental biology of RNA editing in the rat. *J Biol Chem* **265** 12312-12316.

Yao Z., Blackhart B. D., Johnson D. F., Taylor S. M., Haubold K. W. & McCarthy B. J. (1992) Elimination of apolipoprotein B48 formation in rat hepatoma cell lines transfected with mutant human apolipoprotein B cDNA constructs. *J Biol Chem* **267** 1175-1182.

7

RNA EDITING IN PLANT ORGANELLES

J.M. GRIENENBERGER
Institut de Biologie Moléculaire des Plantes, Université Louis Pasteur, 12 rue du Général Zimmer, 67084 Strasbourg Cedex, France

One of the characteristic features of plant cells is the presence of three genomes: those in the nucleus, chloroplasts and mitochondria. Chloroplasts and mitochondria are semi-autonomous organelles that essentially contain all the equipment necessary for the expression of their genes. The products of these genes are generally involved in the formation of large complexes with nuclearly encoded proteins. Therefore, the expression of the three genomes is modulated by complex regulatory processes which can be different in photosynthetic and non-photosynthetic tissues (Douce and Neuburger, 1989).

Many data have been published describing the structure and organization of plant organellar genomes, including the determination of the complete sequence of the chloroplast (cp) genomes of tobacco (Shinozaki et al., 1986), rice (Hiratsuka et al., 1989) and Marchantia polymorpha (Ohyama et al., 1986), and of the mitochondrial (mt) genome of Marchantia polymorpha (Oda et al., 1992).

Much information is available on the transcription and processing of the transcripts of cp genes (Gruissem et al., 1988; Van Grinsven et al., 1988). Much less is known about the same topics for mt transcripts, with only few data on the biochemistry of transcription, RNA polymerase and transcription factors, mRNA stability and maturation. This field is, however, evolving rapidly owing to the recent availability of in vitro transcription and maturation systems (Hanic-Joyce and Gray, 1990, 1991) and the first functional analysis of a plant mitochondrial promoter (Rapp and Stern, 1992).

A number of comprehensive reviews have been published on the molecular biology of chloroplasts and mitochondria, with emphasis on genome structure and organization (Lonsdale, 1989; Newton, 1988), male sterility (Hanson, 1991), transcription and RNA maturation (Gray *et al.*, 1992), and evolution (Gray, 1989), respectively. These organelles have a number of common features. Their genomes appear to be transcribed into either mono- or multi-cistronic RNAs. Some genes are interrupted and the mature transcripts are generated by cis- or trans-splicing. Homologous genes in different species are very well conserved if one considers their primary nucleotide sequence (Palmer, 1990; Palmer *et al.*, 1988).

A further degree of complexity has been added by the discovery of a form of RNA editing in plant mitochondria characterized by frequent C-to-U and occasional U-to-C conversions (Covello *et al.*, 1989; Gualberto *et al.*, 1989; Hiesel *et al.*, 1989). This has recently been followed by the finding that this way of modifying the genetic information is also employed in plant chloroplasts (Hoch *et al.*, 1991; Kudla *et al.*, 1992). Up to now, there is no indication that it is used during the expression of the nuclear genome of plants, but this might change as the number of comparisons between cDNA and the corresponding genes increases. As discussed in Chapter 6, a C-to-U conversion type of RNA editing has been found to occur during the expression of the mammalian apolipoprotein B gene.

The present review will focus on RNA editing in plant mitochondria and chloroplasts. After the description of the discovery of the phenomenon in plant mitochondria, I will discuss when and to what extent plant mtRNAs are edited. Next, I will deal with the consequence of RNA editing for the encoded protein, with the possible mechanisms of the RNA editing process, and with the recent efforts to construct a system *in vitro*. The current knowledge of RNA editing in plant chloroplasts will then be reviewed and, finally, a few thoughts on the evolutionary background of both RNA editing processes will be presented. Excellent reviews have already been published, to which the reader is referred (Bonnard *et al.*, 1992; Gray *et al.*, 1992; Mulligan, 1991; Schuster *et al.*, 1991b; Walbot, 1991).

7.1 RNA EDITING IN PLANT MITOCHONDRIA

7.1.1 Discovery of RNA editing in plant mitochondria
RNA editing in plant mitochondria was discovered simultaneously in three laboratories in 1989 in different ways, as the problems under study were different in each laboratory (Covello and Gray, 1989; Gualberto *et al.*, 1989; Hiesel *et al.*, 1989).

As soon as plant mt gene sequences became available, a number of odd observations indicated that in the plant mt genetic system the universal genetic code is not used. For example, upon comparison of the maize mt cox2 protein sequence with the corresponding bovine and yeast sequences, it was proposed that the plant mt genetic code would not only decode the universal UGG codon as tryptophan, but also the CGG codon (universal arginine codon) (Fox *et al.*,

1981). Other mt genes also contain CGG codons at positions which are conserved as tryptophan in the homologous proteins of other organisms (Hiesel et al., 1983; Gualberto et al., 1988, 1990a). On the other hand, a CGG codon was sometimes found at positions which align with a conserved arginine (Bland et al., 1986; Gualberto et al., 1988, 1990a), and occasionally both situations even occur in the same gene (Gualberto et al., 1990a). We have been trying to identify the specific tRNA which would possess a CCG anticodon but yet specify tryptophan. However, only one tRNA^{Trp} and one tRNA^{Trp} gene were found in bean mitochondria (Maréchal et al., 1985, 1987) and this has been confirmed by the mapping of all tRNA genes in the maize mitochondrial master circle (Sangaré et al., 1990).

The primary sequence of plant mt genes is highly conserved, even though the structure of the corresponding mt genomes can be very different (Palmer and Herbon, 1988). Through the comparison of the deduced protein sequence of cox2 in different plant mitochondria, Gray and coworkers noticed that a significant proportion of the differences observed would be eliminated by C-to-U substitutions. This is the case, for example, for the Arg228 codon in the wheat mt cox2 gene which is at a position conserved as a cysteine in other organisms and is thought to bind a functionally essential copper atom (Covello and Gray, 1989; Covello et al., 1990b). Consequently, the authors proposed that the question of the genetic code in plant mitochondria could be solved by occasional C-to-U conversions.

The first example of edited RNAs appeared in 1987 when Brennicke and coworkers reported differences between the sequence of the Oenothera cox3 gene and the corresponding cDNA, in which two Ts were present instead of Cs (Hiesel and Brennicke, 1987). However, at the time it was not recognized that this was the consequence of RNA editing and these differences were assumed to be cloning artifacts. Clear evidence for the existence of C-to-U conversions by RNA editing in plant mitochondria came when sequences of mtRNAs were compared with the corresponding gene sequences (Covello and Gray, 1989; Gualberto et al., 1989; Lamattina et al., 1989). In these reports artifacts could be ruled out and, simultaneously, re-evaluation of the work on Oenothera indicated that the discrepancies between the cDNA and DNA sequences of cox3 also represented RNA editing events (Hiesel et al., 1989). The occurrence of RNA editing was rapidly verified at the protein level with the determination of the amino acid sequence of wheat atp9 (Bégu et al., 1990; Graves et al., 1990).

These experiments have opened a new field in plant mitochondrial molecular biology, as RNA editing is quite extensive in these organelles. Because the amino acid sequence of a protein can no longer be faithfully deduced from that of the genomic sequence, the knowledge of the RNA sequence is essential to the study of plant mitochondrial genetic expression.

7.1.2 Extent and occurrence of RNA editing

RNA editing is found for all higher plant mt mRNAs for which the sequence has been compared with the corresponding gene sequence. Table 1 shows the number

Table 1 — RNA editing in plant mitochondria

Gene	Plant	Number of editing sites Non-coding			Coding			Number of silent modifications	Altered amino acids		References
		5'	Intron	3'	C-to-U	U-to-C	CC-to-UU		Number	%	
atpA	Oenothera	0	na	0	4	0	0	2	2	0.6	Schuster et al. (1991a)
atp6	Oenothera	0	na	0	21	0	0	1	20	7.1	Schuster & Brennicke (1991a)
	Sorghum	0	na	0	20	0	1	3	16	4.2	Kempken et al. (1991)
atp9	Nicotiana	nd	na	nd	10	0	0	3	7	9.4	Hemould et al. (1992)
	Oenothera	nd	na	nd	4	0	0	0	3	4	Schuster & Brennicke (1990)
	Petunia	nd	na	nd	10	0	0	2	7	9.4	Wintz & Hanson (1991)
	Sorghum	nd	na	nd	8	0	0	1	7	9.4	Salazar et al. (1991)
	Triticum	nd	na	nd	8	0	0	2	5	6.8	Bégu et al. (1990), Nowak & Kück (1990)
cob	Oenothera	0	na	0	14	1	0	1	13	3.3	Schuster et al. (1990a)
cox1	Thuya	nd	na	nd	26	0	2	1	23	nd	Glaubitz & Carlson (1992)
cox2	Oenothera Petunia	0	0	1	22	1	0	6	17	5.8	Hiesel et al. (1990)
	Pisum	2	0	nd	13	1	0	2	12	4.6	Covello & Gray (1990a)
	Triticum	1	0	nd	17	0	1	1	15	5.8	Covello & Gray (1990a)
	Zea	1	0	nd	20	0	1	2	17	6.5	Covello & Gray (1990a), Yang & Mulligan (1991)
cox3	Oenothera	nd	na	6	4	0	0	0	nd	nd	Hiesel et al. (1990)
	Triticum	0	na	0	12	1	0	1	12	4.5	Gualberto et al. (1990b)
nad1	Oenothera	0	2	0	28	0	4	4	20	6	Wissinger et al. (1991)
	Petunia	0	1	0	27	0	4	7	19	6	Conklin et al. (1991)
	Triticum	0	nd	0	17	0	2	1	14	nd	Chapdelaine & Bonen (1991)
nad2	Oenothera	0	4	0	36	0	0	11	29	5.9	Binder et al. (1992)
nad3	Oenothera	0	na	0	16	0	2	2	12	10.1	Schuster et al.(1991b)
	Triticum	0	na	0	21	0	4	1	16	13.5	Gualberto et al. (1991)
nad4	Triticum	0	nd	0	23	0	1	0	22	4.5	Lamattina et al. (1991), Lamattina et al. (1989)
nad5	Arabidopsis	0	nd	0	8	0	0	1	nd	nd	Knoops et al. (1991)
	Oenothera	0	3	0	26	0	0	0	26	3.9	Knoops et al. (1991)
	Triticum	0	0	0	11	0	1	0	10	1.5	Desouza et al. (1991)
nad6	Triticum	0	na	0	15	0	4	4	9	3.6	Haouazine et al. (1992)
rpsl2	Triticum	0	na	0	6	0	0	0	6	4.8	Gualberto et al. (1990b)
rpsl3	Daucus	0	na	0	4	0	0	1	3	2.6	Wissinger et al. (1990)
	oenothera	0	na	0	3	0	0	1	2	1.7	Wissinger et al. (1990)
rpsl4	Oenothera	1	na	0	2	0	0	0	2	2	Schuster et al. (1991a)
rpsl9	Oenothera	0	na	0	7	0	0	2	5	5.3	Schuster et al. (1991b)
	Petunia	0	na	0	7	0	2	1	4	4.2	Conklin et al. (1991a)
orfB	Oenothera	0	na	2	6	0	1	1	5	3.1	Schuster et al. (1991c)
orf156	Triticum	0	na	0	4	0	0	1	3	1.9	Gualberto et al. (1990b)
orf221	Zea	0	na	0	10	0	0	2	8	3.6	Ward & Levings (1991)
T-urf3	Zea	0	na	0	0	0	0	0	0	0	Ward & Levings (1991)

The names of the genes are written according to the plant mitochondrial nomenclature as defined by Lonsdale and Leaver (1988).
nd: not determined, na: not applicable

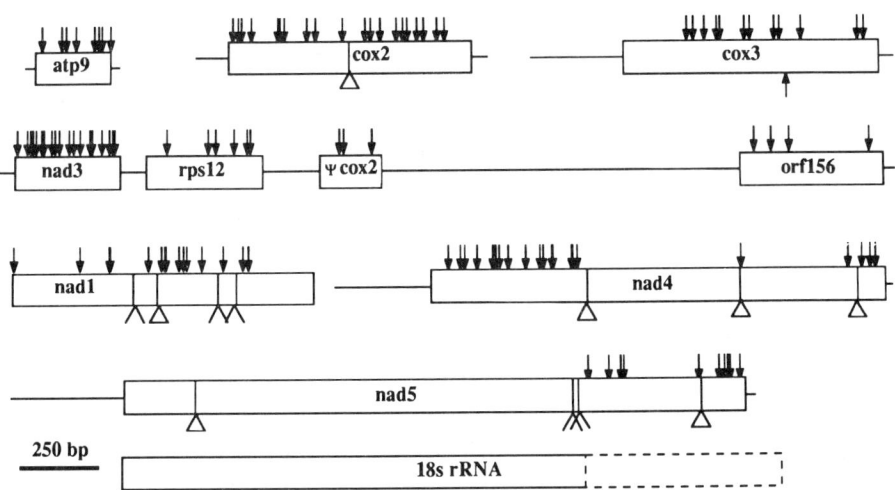

Fig. 1 — Distribution of editing sites in wheat mt transcripts. The position of editing sites is indicated by arrows. Coding sequences are indicated by boxes, non coding segments by lines. The position of introns is indicated by Λ (trans-spliced introns) or Δ (cis-spliced introns).

of edited nucleotides reported so far for mitochondrial transcripts for which sufficient data are available. In addition to the contents of Table 1, some incomplete data have been reported for the following transcripts: cox1 and cob of wheat (Gualberto *et al.*, 1989), mat-r (maturase-related) of *Oenothera* (Wissinger *et al.*, 1991) and rps13 of maize (Hunt and Newton, 1991). The distribution of the editing sites over a given RNA is random for most transcripts (see Fig. 1). Wheat nad5 RNA, however, shows a very unequal distribution, all sites being clustered in exons 4 and 5.

These results are derived from different plants, monocots such as wheat, maize and *Sorghum* as well as dicots such as *Oenothera, Arabodopsis, Petunia,* pea, tobacco and carrot. The best-studied plants are wheat and *Oenothera* for which 134 and 177 editing sites, respectively, have been identified thusfar. RNA editing has recently been found in western red cedar (*Thuya plicata*), a gymnosperm (Glaubitz and Carlson, 1992). It probably does not occur in the unicellular green algae *Chlamydomonas reinhardtii* (Boer and Gray, 1986a, b, 1988) and in the bryophyte *Marchantia polymorpha* (Oda *et al.*, 1992; Ohyama *et al.*, 1991), see Section 7.4.

The vast majority of editing events (474 out of 478) involve the conversion of a C (encoded in the template DNA) into a U. The reverse modification, the conversion of a DNA-encoded U into a C, has been found in four instances (< 1%). The U-to-C reverse modification was first reported for the cob transcript of *Oenothera* (Schuster *et al.*, 1990a). In the case of wheat atp9, a U-to-C modification was found in only one cDNA clone out of fifty (Bégu et al., 1990) and was not detected by another group (Nowak et al., 1990). The U-to-C modification found in the *Oenothera* cox2 transcript appears in one clone out of three (Hiesel et al., 1990). However, in the case of the wheat cox3 transcript, the reverse modification was found in the 12 different cDNA clones sequenced (Gualberto et al., 1990b) and this is a good indication that the U-to-C change is a real editing event. The presence of a T in the mt genome was checked using direct sequencing of the uncloned genomic mtDNA in order to eliminate cloning artifacts. It is not clear whether reverse editing should be considered as an error of the editing machinery or as rare but *bona fide* editing events. In this case, analysis of the U-to-C conversions would be quite important for the comprehension of the biochemistry of editing.

Editing sites along transcripts are almost always located in the coding regions of the RNA, e.g. in Table 1 for 478 editing sites reported in coding regions, only 5 editing sites have been found in the upstream and 7 in the downstream non-coding regions. The effect of RNA editing in these non-translated regions is not yet understood. One edited nucleotide reported in the upstream region of the *Oenothera* rpsl4 transcript can improve a putative ribosome binding site (Schuster et al., 1990b). However, the existence of such binding sites has not been firmly established in plant mitochondria (Boer et al., 1985). Another interesting editing site is found in a transcribed repeated sequence which is located in the upstream non-coding region of at least two wheat mt RNAs, cox2 and orf25 (Bonen et al., 1990; Covello and Gray, 1989). One can imagine that these editing sites modify the primary sequence or even the secondary structure of regulatory elements such as ribosome binding sites as well as structures involved in the stability of the mRNA (Schuster et al., 1986). The only protein-coding region that does not seem to be edited is that found in the Turf13 transcript. This RNA is derived from a chimeric gene which is generated by recombination events in the mitochondria of the male sterile line T of maize (Ward and Levings, 1991). This exception can be explained by the realization that the Turf13 gene is mainly assembled from sequences homologous to the 26S ribosomal RNA gene region (Dewey et al., 1986). Structural RNAs are largely unedited in plant mitochondria (see below).

Introns have been found in several plant mitochondrial genes, such as cox2 (Bonen et al., 1984; Fox and Leaver, 1981; Gwynn et al., 1987; Kao et al., 1984; Pruitt and Hanson, 1989), nadl, (Chapdelaine and Bonen, 1991; Wissinger et al., 1991), nad2, (Binder et al., 1992), nad4 (Lamattina and Grienenberger, 1991; Lamattina et al., 1989) and nad5 (Desouza et al., 1991; Knoop et al., 1991). Plant mitochondrial introns clearly belong to class II introns (Michel et al., 1989) and they are spliced via cis- as well as via trans-splicing. Table 1 shows that 10 editing sites have been identified either in cis- or trans-splicing introns in *Petunia* (Conklin et al., 1991) or in *Oenothera* (Binder et al., 1992; Knoop et al., 1991; Wissinger et al., 1991). Where modeling of the intron is possible, it appears that some of the editing events modify the secondary structure of some well conserved domains of group II introns and may be crucial for correct or efficient splicing (see Wissinger *et al.*, 1992).

The postulated mechanism of splicing for class II introns involves specific folding of the unspliced RNA molecules through the recognition of two sequences of the upstream exon (IBS1, IBS2) by two sequences located within the intron (EBS1, EBS2) (Michel *et al.*, 1989). Correct folding of the intron could therefore be affected by editing events occurring in IBS1 or IBS2. This is the case for the splicing of wheat and maize cox2 introns whose secondary structure and EBS1 and EBS2 sequences have been identified (Michel *et al.*, 1989). Editing occurs at codon 87, altering the sequence of IBS1 (AUCGGA) in both wheat and maize. The sequence of the complementary EBS1 is UCCGAU. The nucleotide pair obtained during folding of the intron is either G:C for the non-edited nucleotide or G:U for the edited one. Both pairs are allowed in double-stranded RNA structures as often found in tRNAs, rRNAs and (pre)mRNAs. Mutational analysis of the group I intron of the *Tetrahymena* pre-ribosomal RNA has shown that some G:U pairs are essential for proper splicing to occur and cannot be replaced by G:C basepairs (Price *et al.*, 1985). It is unclear whether the status of the wheat and maize cox2 IBS1 nucleotide (edited or unedited) is relevant for splicing of this intron, but the possibility remains that the change in stability of the base paired region upon editing is one of the requirements for proper splicing to occur. However, this assumption has to be demonstrated using splicing systems *in vitro*.

Structural RNAs do not seem to be edited. The study of the 18S mitochondrial rRNA in wheat has shown that neither the 200 nucleotides at the 3' end, as determined by direct sequencing (Spencer *et al.*, 1984), nor the sequence of six different cDNA clones covering 80% of the mature rRNA (our unpublished results) contain any edited nucleotides. No editing was found in the 5S rRNA and the 1820 nucleotides analyzed of the 18S rRNA of *Oenothera*. However, one U-to-C and one C-to-U editing site have been found in one partial cDNA clone of the 26S rRNA of *Oenothera* (Schuster *et al.*, 1991a). The use of RNA editing in structural RNA in plant mitochondria remains to be ascertained by further examples.

In view of the high frequency of edited nucleotides in mRNA but their very rare occurrence in structural RNA, it is possible to use the presence of edited nucleotides in an RNA as a criterion to decide whether an open reading frame represents a protein gene, as shown for orf156 in wheat mitochondria (Gualberto *et al.*, 1991). Amino acid sequences of plant mitochondrial proteins are not available except for subunit 9 of wheat ATPase (Bégu *et al.*, 1990; Graves *et al.*, 1990). Eight editing positions are found in the wheat atp9 transcript, five of them inducing a change of amino acid and one creating a stop codon. The protein sequence of the first 32 residues (Graves *et al.*, 1990) shows, at positions 7 and 28, a leucine and a phenylalanine corresponding to the edited codons, UUA and UUC respectively. Residues corresponding to non-edited codons could not be detected. Carboxy-terminal sequence data confirm the change from serine to leucine at position 71 as well as the shortened size of the protein due to the creation of a stop codon at the expected position. This strongly indicates that, in plant mitochondria, proteins are produced from edited mRNAs.

7.1.3 Partial editing

In the very first report on plant mitochondrial RNA editing, the possibility of partially edited RNAs was deduced from an ambiguity, i.e. the presence at the same position of a C and a U, in an RNA sequence of wheat nad3 at one editing position (Gualberto *et al.*, 1989). Partial RNA editing can be defined as the presence in the pool of mRNA of some transcripts that are not edited at all known editing sites. The existence of partially edited transcripts was first clearly established by sequencing eight cDNA clones of *Oenothera* nad3 transcripts (Schuster *et al.*, 1990c). Partially edited transcripts have been studied in order to see whether they could be intermediates in the process of editing and/or code for different polypeptides.

Partially edited transcripts have now been reported for a number of genes, such as cox2 of *Oenothera*, (Hiesel *et al.*, 1990) wheat and maize (Covello and Gray, 1990a; Yang and Mulligan, 1991), nad3 and rpsl2 of wheat (Gualberto *et al.*, 1991) and *Oenothera* (Schuster *et al.*, 1990c), rpsl3 of *Oenothera*, (Wissinger *et al.*, 1990), rps19 of *Petunia* (Conklin and Hanson, 1991), rps3 of maize (Hunt and Newton, 1991), and nad1 of wheat (Chapdelaine and Bonen, 1991) and *Oenothera* (Wissinger *et al.*, 1991). On the other hand, only fully edited transcripts are found for some other genes, such as cox3 of wheat (Gualberto *et al.*, 1990b), atp9 of wheat (Nowak and Kück, 1990) and *Oenothera* (Schuster and Brennicke, 1990), nad4 (Lamattina and Grienenberger, 1991) and orfl56 (Gualberto *et al.*, 1991) of wheat, and rpsl4 (Schuster *et al.*, 1990b) in *Oenothera*.

Theoretically, all partially edited transcripts could be translated because the C-to-U conversions do not generally alter the reading frame of a gene. It has been proposed that differential editing could be tolerated if it occurs at positions where the modification is silent or corresponds to a non-conserved amino acid (Schuster and Brennicke, 1990; Schuster *et al.*, 1990c). Nevertheless, as the number of different partially edited cDNA clones increases, it seems essential that editing proceeds efficiently to completion before translation. If not, a virtually endless number of different proteins would be synthesized. This would imply that these partially edited transcripts are intermediates in the editing process as postulated by Gualberto *et al.* (1991).

Transcription in higher plant mitochondria is a highly complex process, giving rise to multiple RNA molecules of different sizes from the same gene. These could be produced by differential processing or be derived from different transcription initiation events (Mulligan *et al.*, 1991). There appears to be a correlation between the occurrence of such multiple RNAs and that of partially edited sequences. In the case of rapidly processed mRNAs, as in wheat where usually only one major transcript is detected, editing intermediates represent a low percentage of the transcripts and a high number of cDNA clones has to be generated in order to find them. For most other species, multiple transcripts are detected on Northern blots. In *Oenothera*, nad3 is transcribed into up to six different RNA molecules ranging in size from 1200 to 7500 nt which are found in comparable amounts (Schuster *et al.*, 1990c). The transcripts that are most likely processing intermediates show partial editing. Processing intermediates are easily detected for RNAs of maize, *Petunia* and *Oenothera* where the existence of partially edited transcripts was reported (Conklin and Hanson, 1991; Covello and Gray, 1990a; Hiesel *et al.*, 1990; Hunt and Newton, 1991; Schuster *et al.*, 1990c; Wissinger *et al.*, 1990; Wissinger *et al.*, 1991).

The cox2 genes of *Petunia* (Sutton *et al.*, 1991) and maize (Yang and Mulligan, 1991) contain an intron. cDNA clones of unspliced transcripts of these two species showed a full range of editing from 0% to 100%. However, incompletely edited transcripts are prevalent, whereas spliced mature transcripts are almost completely edited. Editing of *Petunia* nad1 pre-mRNA, on the other hand, can occur before cis- and trans-splicing (Sutton *et al.*, 1991).

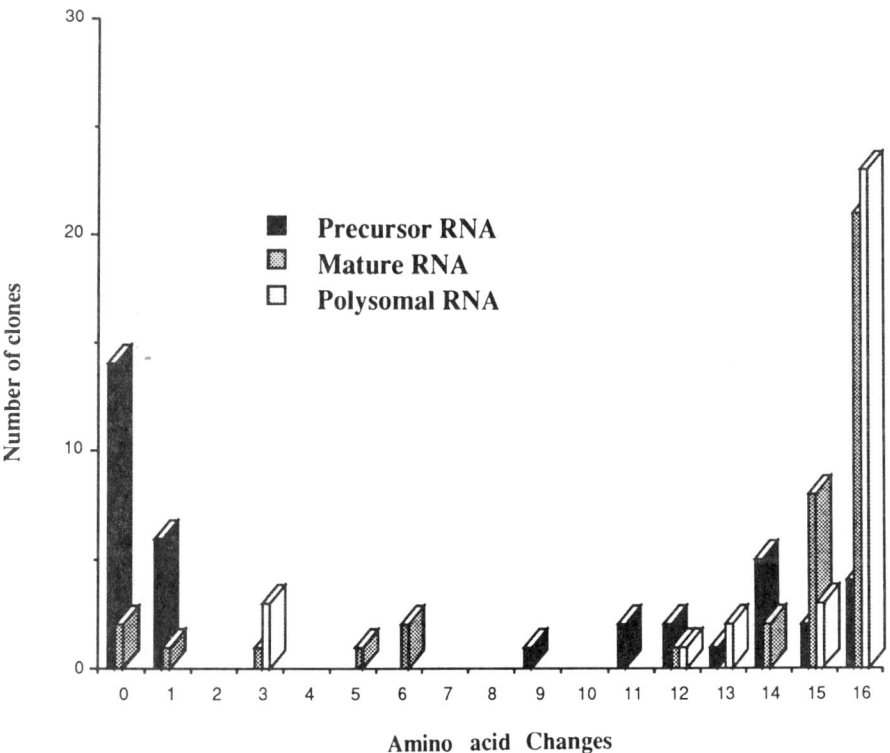

Fig. 2 — Effect of partial editing on the deduced nad3 protein sequence. Editing of nad3 results in a change of 16 amino acids. The bars indicate the number of editing events observed in cDNAs clones obtained from the 2900 nt transcript from total mt RNA (precursor RNA), the 900 nt transcript from total mtRNA (Mature RNA) and the 900 nt transcript from polysomal enriched mRNA (Polysomal RNA). Data are extracted from Gualberto *et al.* (1991).

In wheat, the nad3 and rps12 genes (Gualberto *et al.*, 1991) are transcribed into a precursor RNA of 2900 nt and a mature RNA of 900 nt whose editing state could be studied separately depending on the position of the primers used for amplification. Most of the 2900 nt cDNA clones correspond to non-edited and partially edited transcripts (Fig. 2). The existence of unedited precursor RNA strongly indicates that RNA editing is a posttranscriptional process. The distribution of the edited nucleotides along the partially edited precursor RNAs is essentially random, i.e. without the evident 3' to 5' polarity found in trypanosome

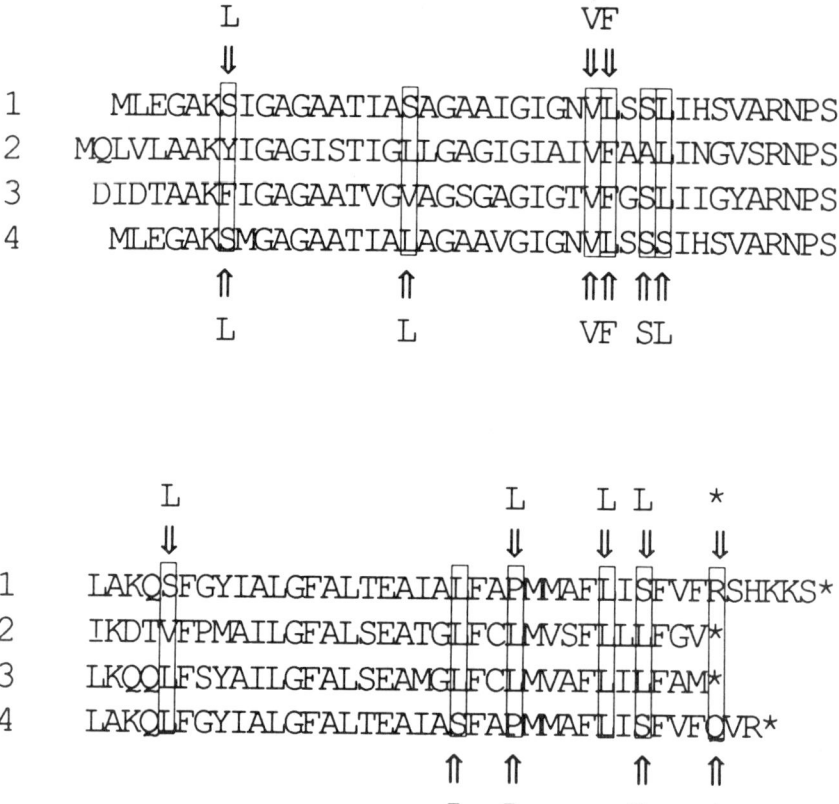

Fig. 3 — RNA editing increases the conservation of plant mt protein sequences. Amino acid comparison of the wheat and *Petunia* atp9 polypeptide sequences deduced from the DNA and edited mRNA sequences with the respective atp9 protein of yeast (mitochondrially encoded) and bovine (nuclearly encoded). The amino acids which are specified by an edited codon are shown in the top line for wheat and the bottom line for *Petunia* as indicated by arrows. 1 wheat, (Bégu *et al.*, 1990); 2 yeast, (Hensgens *et al.*, 1979); 3 bovine, (Gay and Walker, 1985); 4 *Petunia*, (Wintz and Hanson, 1991).

kinetoplasts (Simpson and Shaw, 1989; see Chapters 2 and 3). In contrast to the precursor cDNA clones, most of the cDNAs derived from mature RNAs are completely or almost completely edited (Fig. 2; Gualberto *et al.*, 1991). Only a few cDNAs show no editing at all. Nearly all transcripts are fully edited in polysome-enriched RNA. Similar results were found with cDNA clones obtained from total maize cox2 polysomal RNA (Yang and Mulligan, 1991). It is therefore very likely that one single protein is produced. These results indicate that, in general, RNA editing in plant mitochondria occurs in parallel with maturation processes such as splicing. There does not appear to be a strict sequence of events, though, since in almost all cases unprocessed precursor RNAs are found that are partially or even completely edited.

7.1.4 The protein sequence after editing

Most of the editing events (86%) induce a change in the encoded amino acid, affecting the first position (49%) or the second position (30%) of the codon (see also   Table 3 of Chapter 6). The remaining 7% are adjacent editing events leading to the conversion of a proline codon (CCN) into leucine (UUR) or phenylalanine (UUY) codons. A minority of editing sites (14%) have been found at the wobble position of the codon and have therefore no influence on the encoded amino acid.

RNA editing can create initiation and stop codons. The modification of a threonine codon ACG into the methionine codon AUG was found in cox2 genes of wheat, maize, and pea at different internal positions (Covello and Gray, 1990a). In the case of wheat nad1, where no AUG could be found at the position of the predicted start codon, the initiation codon is created by RNA editing (Chapdelaine and Bonen, 1991). In *Oenothera* nad1 the same modification occurs six codons after the first in-frame AUG (Wissinger *et al.*, 1991). The AUG created by editing would initiate the translation at the same position in both species and is therefore very likely the real initiation codon.

Stop codons could be created by editing of glutamine codons (CAG, CAA) and arginine codons (CGA) (Fig. 3). The creation of UGA was reported for atp9 RNA of wheat (Bégu *et al.*, 1990; Nowak and Kück, 1990) and *Oenothera* (Schuster and Brennicke, 1990) and the creation of UAA for atp9 RNA of *Petunia* (Wintz and Hanson, 1991). In the case of wheat, this modification was confirmed by protein sequencing (Bégu *et al.*, 1990).

Plant mitochondrial RNA editing was first described as a correction mechanism which results in the conservation of protein sequences (Covello and Gray, 1989; Gualberto *et al.*, 1989). This better conservation becomes clear when one compares protein sequences of different plant species with the homologous proteins of non-plant organisms (Fig. 3). The discovery of RNA editing therefore solved not only the problems with the plant mitochondrial genetic code as discussed above, but also further confirmed existing ideas about the function of conserved amino acids. This is well illustrated by the study of the Cu_A-binding site of the wheat cytochrome c oxidase subunit 2 (Covello and Gray, 1990b). After editing, amino acid residue 228 is changed into a conserved Cys, supporting its role in Cu-binding. As a further consequence of editing, a Met, conserved in the cox2 gene of all species except plants, appears at position 235 of all cox2 transcripts in higher plants. This firmly established the evolutionary conservation of this methionine (corresponding to Met 207 in bovine cox2) and led to a new model for the structure of the Cu_A-binding site (Covello and Gray, 1990b). Taking RNA editing into account, most mitochondrial proteins appear nearly identical in all plant species. However, some differences remain, including amino acids corresponding to edited positions. When RNA editing occurs at non-conserved positions, it could be an indication of a species-specific role of the corresponding amino acid such as interaction between subunits of the respiratory chain complexes.

7.2 RNA EDITING IN CHLOROPLASTS

The complete sequence of chloroplastic DNA has been reported for a number of species, both from higher (Hiratsuka et al., 1989; Shinozaki et al., 1986) and from lower (Ohyama et al., 1988) plants. Apart from structural differences in the organization of the circular cp genomes, such as inversions, the presence and size of inverted repeats, etc., the nucleotide sequences of the identified or unidentified reading frames are very similar and well conserved, allowing easy comparison between species.

In such a comparison, the group of H. Kössel detected some differences which appeared essential for the expression of at least two genes. The maize cp rpl2 gene (Hoch et al., 1991) and the tobacco cp psbL gene (a subunit of photosystem 2) (Kudla et al., 1992) contain an ACG codon where the other cp genomes have an initiation codon ATG. Because the ACG (threonine) codon would not allow the expression of these genes, they postulated that C-to-U conversions, analogous to RNA editing in plant mitochondria, create the correct reading frames. Analysis of the corresponding cDNAs showed that this actually is the case, which demonstrated that RNA editing is also used in the expression of the chloroplast genome.

An analogous experiment has been carried out to identify the initiation codon of psbL in a non-photosynthetic organelle, the bell pepper (Capsicum annuum) chromoplast (Kuntz et al., 1992). The same situation as in tobacco has been found (an ACG codon in the gene and an AUG codon in the mRNA), indicating that RNA editing is not involved in the regulation of expression of photosynthetic genes during the differentiation of the chloroplast into chromoplast.

More recently, (Maier et al., 1992), found four other editing sites in the coding region of maize cp ndhA, a gene which is supposed to code for the chloroplast equivalent of subunit 1 of the mitochondrial NADH dehydrogenase (complex 1). All these editing events are C-to-U conversions and all of them modify the identity of the specified amino acid. Each of the six editing sites identified in cp mRNA result in the restoration of codons for conserved amino acids, enhancing the similarity with the homologous chloroplast proteins of different plants. Two of these editing sites are located in highly conserved regions of the equivalent proteins in most of the mitochondria (plant, fungi,

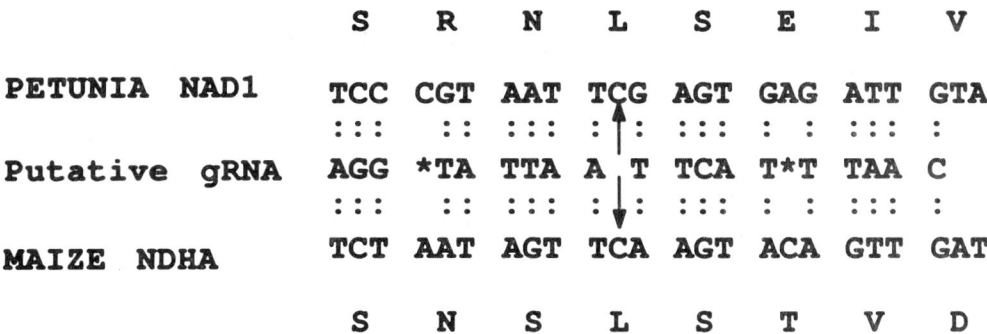

Fig. 4. —— Comparison between the nucleotide sequences of *Petunia* mt nad1 (Conclin *et al.*, 1991) and of maize cp ndhA (Maier *et al.*, 1992) around a homologous editing site. The arrow indicates the common editing site. A putative gRNA is shown allowing G:U base pairs.

mammals) and restore a highly conserved leucine. Even more interestingly, one of these editing sites is located at exactly the same position in the transcript of the mitochondrial homolog of the ndh A gene, the nad 1 gene (see Fig. 4) (Chapdelaine and Bonen, 1991; Conklin *et al.*, 1991; Wissinger *et al.*, 1991). Even though the ndhA protein has not yet been identified in chloroplast, this example of conserved editing might suggest co-evolution of RNA editing in both organelles.

7.3 BIOCHEMISTRY OF RNA EDITING IN PLANT ORGANELLES

Most of the RNA editing in plant mitochondria and chloroplasts is the conversion of a C encoded in the DNA into a U in the RNA. Even if the number of editing sites in chloroplast RNA is quite small compared with that in mitochondria, the biochemistry of these modifications may be very similar and the same assumptions can be made. RNA editing in plant organelles includes at least two steps: (i) selection of the specific cytosines to be edited and (ii) their biochemical modification. The specificity of site selection could reside in a common sequence motif to be identified by the enzymatic machinery. In such a scheme, site selection would be similar to that for the edited C of the apolipoprotein B mRNA, as discussed in the previous chapter. Given the high frequency of Cs to be edited, one could also envisage that the specificity is provided by specific guides or templates likely to be RNA, as in trypanosomal RNA editing (Chapters 2 and 3). Nothing is known yet, however, about the nature of the (enzymatic) factors involved in editing in plant mitochondria, which could be single proteins, RNAs or complex ribonucleoprotein particles. Some clues as to their identity and how they might select an editing site are discussed in the next sections.

7.3.1 Site selection

In wheat mitochondria, 134 editing sites have been identified up to now. What are the factors involved in the specificity of the process? A first clue comes from the analysis of the editing of part of the transcript coding for nad3 and rpsl2. Higher plant mitochondrial genomes contain a number of repeated sequences which consist of parts of known genes (Bonen and Bird, 1988). In wheat, the 3' end sequence of the nad3/rps12 primary transcript contains a 193 nucleotide segment which is identical to part of the first exon of cox2 (Gualberto *et al.*, 1991). The surrounding sequences are not edited, but three modifications occur in the cox2 insertion at the known editing sites of the genuine cox2 transcript. This indicates that a short RNA segment is sufficient to specify an editing site and that long distance tertiary structure interactions, most likely, do not play a role. The local sequence must contain the main determinants of the specificity.

No clear consensus sequence can be identified from the primary nucleotide sequences surrounding edited cytidines, although deviations from mean nucleotide frequencies exist. The nucleotide located immediately upstream of the edited C is most often a pyrimidine (mainly a uridine) and very rarely a guanosine. A high frequency (43%) of an UCR motif was found when looking at 61 non-homologous editing sites (Covello and Gray, 1990a). The significance of these observations is unknown still, but it is very unlikely that the editing specificity relies on such a small and loose motif which is also present at numerous positions that are not edited. Moreover, it is also very unlikely that each site is recognized by separate protein factor(s), in which case it would be necessary to postulate the existence of hundreds (thousands) of different editing proteins.

It is attractive to assume, therefore, that the specificity is somehow provided by nucleic acids. For example, base pairing interactions could provide the information *in cis*, as recently suggested for the unique U-to-C RNA editing event modifying the genomic RNA of human hepatitis delta virus (Zheng *et al.*, 1992). Inspection of the available sequences, however, has failed to show evidence for such secondary-structure determinants. Alternatively, trans-acting antisense guide RNAs (gRNAs) similar to those identified in trypanosomes (Blum and Simpson, 1990; see Chapters 2 and 3) could play a role.

The sequences surrounding a number of editing sites can be grouped into families with a certain degree of sequence conservation (Gualberto *et al.*, 1990b, 1991). If gRNAs are indeed used, each of these groups could base pair to a putative antisense gRNA in RNA/RNA interactions which include G:U base pairing. On the other hand, one could imagine a separate gRNA for each site (or cluster of sites since editing sites can be close to each other) in one transcript. Templates or guides could also consist of full-length antisense RNA carrying the edited sequence and able to direct the editing process. Their existence, however, remains to be demonstrated.

An interesting case is provided by the editing site discussed above, which is homologous in the mitochondrial nad1 and chloroplast ndhA transcripts (Maier *et al.*, 1992). Fig. 4 shows the surrounding sequence of the editing site in both transcripts and it is clear that these sequences are quite similar and could therefore possibly basepair to a common (putative) gRNA. This raises the interesting but of course highly speculative possibility that the same guide RNA or a very similar one is used for the editing of a chloroplast and a mitochondrial transcript.

7.3.2 Mechanism of the nucleotide conversions

Since the evidence for the conversion of cytidine into uridine is derived only from cDNA sequence analysis, one cannot be sure that editing really produces Us and not some unusual modified nucleotide which also directs the incorporation of an A by reverse transcriptase and is recognized as a U by a tRNA during protein synthesis. The identification of the edited nucleotide as a U has recently been reported for the editing of human apolipoprotein B RNA (Hodges *et al.*, 1991; see Chapter 6) but remains to be done for plant mitochondrial editing because it requires the availability of an *in vitro* editing system.

Assuming the edited nucleotide is indeed a U, the simplest mechanism for C-to-U conversion would be deamination at position 4 of cytosine, leading to a uracil residue; see Fig. 7 of Chapter 6. This could be achieved by cytidine deaminase, an enzyme operative in nucleotide metabolism (O'Donovan and Neuhard, 1970). The reverse U-to-C modification found in plant mitochondria could be carried out by a CTP synthetase in an ATP-dependent reaction (Zalkin, 1985), since it is difficult to envisage that it is done by the reverse action of the cytidine deaminase. This would imply the presence of two different enzymatic activities. Interestingly, it has recently been proposed that the human hepatitis delta virus U-to-C change mentioned above is carried out by a CTP synthetase-like enzyme (Taylor *et al.*, 1992).

Another possibility would be exchange of the base without cleavage of the sugar-phosphate backbone. Such trans-glycosylation reactions have been described in transfer RNAs (Björk *et al.*, 1987; Okada and Nishimura, 1979). Pyrimidine replacement is also found in DNA repair mechanisms (Olsen *et al.*, 1989). This process, like deamination, does not need any break in the RNA chain.

The third possible mechanism involves cleavage of the RNA chain, deletion of a cytidine, addition of an uracil followed by religation. The nucleotide exchange would involve several steps similar to those described for U additions and deletions in editing of mtRNAs of trypanosomes (Blum *et al.*, 1990; Cech, 1991; Chapters 2 and 3).

In order to decide between these possible mechanisms an *in vitro* RNA editing system should be available. Numerous attempts are currently being made to design such a system. They are generally based on a primer extension test similar to that described for the *in vitro* apolipoprotein B system (Driscoll *et al.*, 1989; Fig. 2 of Chapter 6). So far, one report exists of editing in plant mitochondria *in vitro* (Araya *et al.*, 1992). The formation of edited atp9 RNAs as a result of incubation of unedited precursor RNA with the soluble fraction of purified wheat mitochondria was checked by sequence analysis of cloned PCR products. The sensitivity of the activity to high temperatures and protease digestion argues in favor of the involvement of proteins, and a micrococcal nuclease pre-treatment of the extract which abolishes the RNA editing activity suggests that nucleic acids may also be involved. Such an assay system can be used as a starting point for further analysis of the components involved in RNA editing in plant mitochondria and their mode of action, but it is too early to make any definitive statements on the precise mechanism (base conversion, base excision, or nucleotide deletion/ insertion) that is used (see also Wissinger *et al.*, 1992). It would be unexpected, however, if at least some aspects of the pathway followed do not resemble those of the C→U conversion reaction described in the previous chapter, since it is difficult to envisage that two completely different ways of converting Cs of an RNA into Us have arisen in nature (see also Section 7.4).

7.4 EVOLUTION OF RNA EDITING IN PLANTS

Although the pyrimidine-conversion type of RNA editing is now a well-established phenomenon in higher plants, the question is whether it is present all along the plant kingdom. The only information available in this respect comes from the lower plant *Marchantia polymorpha*. Comparison of the cox2 DNA sequence of *Marchantia* with other cox2 genes strongly suggests that RNA editing does not occur in *Marchantia* (Ohyama *et al.*, 1991). This was confirmed by the publication of the sequence of the complete mt genome of *Marchantia polymorpha* (Oda *et al.*, 1992), which shows that there is no need of RNA editing to synthesize proteins with high amino acid similarity to their homologous counterparts. At all editing sites of higher plant mitochondrial gene sequences (where a C is present), there is a T in the corresponding *Marchantia* genes. There are two possibilities to explain this observation: (1) pyrimidine conversion appeared after the divergence of angiosperm plants from bryophytes or (2) it is an ancient mechanism which has been lost in *Marchantia*. The first option could imply that this type of RNA editing has evolved to correct mutations which arose in the mitochondrial DNA. It is difficult to envisage, however, that evolution has created this mechanism *de novo* in a slowly evolving genome. Furthermore, these mutations should not be restricted to one type of nucleotide conversion.

The mt type of RNA editing is also found in chloroplasts, albeit to a lower extent. This might indeed suggest that this form of RNA editing is an ancient phenomenon originating from the very beginning of life. In this view it could be a characteristic of the prokaryotes that were the evolutionary ancestors of chloroplasts and mitochondria and may even have been present in some of their endosymbiont hosts, given the fact that a C-to-U conversion type of editing is also found in a nuclearly encoded RNA (Chapter 6). This assumption implies that plant mitochondria may not have the same phylogenetic origin as mitochondria that do not edit their RNAs or use a mechanistically different type of editing. Such a scenario would be in good agreement with the possible polyphyletic origin of mitochondria as proposed by (Gray *et al.*, 1989). After the initial endosymbiotic events, the different rates of nucleotide mutation, much quicker in chloroplasts than in mitochondria (Palmer and Herbon, 1988), could explain why RNA editing is much more important in mitochondria than in chloroplasts, the higher rate of mutation in chloroplast DNA resulting in a faster loss of editing sites.

A possible clue as to how such a loss of editing sites might have occurred is provided by the observation that the genetic information that flows from the mitochondrion to the nucleus in plants is in the edited form. For example, most legumes have cox2 sequences in both the nucleus and the mitochondria, but the cox2 gene is absent from the mitochondrial genome of two species of the genus *Vigna*, *V. unguiculata* (cowpea) and *V. radiata* (mungbean). Since cox2 is encoded in the mitochondria in the closely related genus *Phaseolus*, the loss might be a recent event (Nugent and Palmer, 1991). The nuclear *Vigna* cox2 genes have the 'edited' sequence, and therefore the transfer of information from the mitochondria to the nucleus is likely to have occurred through an edited RNA intermediate, probably after reverse transcription. This transfer could have been achieved in several ways depending on where the reverse transcriptase step occurred.

A similar mechanism can be put forward to explain the loss of editing sites inside the organelles, either chloroplasts or mitochondria. An edited RNA or RNA segment could have been reversely transcribed and the cDNA inserted in the mt or cp genomes by homologous recombination, leading to the creation of genes containing the edited sequence. This could explain the large difference, in the extent of editing of some genes between different species. As an example, atp6 is edited 21 and 20 times in *Oenothera* (Schuster *et al.*, 1991a) and *Sorghum* (Kempken *et al.*, 1991), respectively, but only once in rapeseed (Handa and Nakajima, 1992) (see Table 1).

7.5 CONCLUDING REMARKS

The discovery of RNA editing has resulted in an extra addition to the already extensive list of processes required for gene expression, such as transcription, RNA splicing, translation and posttranslational protein modification. In this chapter, the consequences of RNA editing for the genetic flow from DNA to protein in plant organelles have been discussed.

RNA editing can activate open reading frames by the creation of initiation codons, can greatly modify the genetic message, and can reduce the size of the encoded protein by the creation of stop codons. Therefore, it is no longer sufficient to describe a DNA sequence and compare it with the existing sequence data banks to be able to identify and discuss the encoded protein. The real sequence to be considered is in fact that of the mRNA (or more easily that of the corresponding cDNA). These considerations have also practical consequences, such as for example the necessity to use the edited sequence when synthesizing mitochondrial or chloroplast proteins in heterologous expression systems, for instance in *E. coli*.

Because it seems that translation of a transcript is dependent on completion of editing, it is likely that this step of the genetic expression of plant organelles will be of the utmost importance for control and regulation. It will be best to keep these controls in mind when devising systems and methods for the transformation of organelles, an obvious and important step towards plant organelle biotechnology.

Finally, and in my view more importantly, the study of RNA editing in plant organelles and its occurrence during evolution, as compared with the related phenomena in trypanosomes, slime mold, and mammals, could well provide a probe for our understanding of the very first events of the onset of life.

ACKNOWLEDGEMENTS

The author wishes to thank all his colleagues from IBMP-CNRS in Strasbourg (France) who have been working on RNA editing in plant mitochondria for their contributions and many helpful discussions: Dr G. Bonnard, Dr D. Gonzalez, Dr J. M. Gualberto, Dr L. Lamattina and Dr H. van der Spek. The work described in this review was supported by the Centre National de la Recherche Scientifique, Paris, France.

REFERENCES

Araya, A., Domec, C., Bégu, D. & Litvak, S. (1992) An *in vitro* system for the editing of ATP synthase subunit-9 messenger RNA using wheat mitochondrial extracts. *Proc. Natl. Acad. Sci. USA* **89** 1040-1044.

Bégu, D., Graves, P.V., Domec, C., Arselin, G., Litvak, S. & Araya, A. (1990) RNA editing of wheat mitochondrial ATP synthase subunit 9: direct protein and cDNA sequencing. *Plant Cell* **2** 1283-1290.

Binder S., Marchfelder, A., Brennicke, A. & Wissinger, B. (1992) RNA editing in trans-splicing intron sequences of nad2 messenger RNAs in *Oenothera* mitochondria. *J. Biol. Chem.* **267** 7615-7623.

Björk, G.R., Ericson, J.U., Gustafsson, C.D.E., Hagervall, T.D., Jonsson, Y.H. & Wikstrom, P.M. (1987) Transfer RNA modification. *Annu. Rev. Biochem.* **56** 263-287.

Bland, M.M., Levings III, C.S. & Matzinger, D.F. (1986) The tobacco mitochondrial ATPase subunit 9 gene is closely linked to an open reading frame for a ribosomal protein. *Mol. Gen. Genet.* **204** 8-16.

Blum, B., Balakara, N. & Simpson, L. (1990) A model for RNA editing in kinetoplastid mitochondria: small "guide RNA" molecules transcribed from maxicircle DNA provide the edited sequence information. *Cell* **60** 189-198.

Blum, B. & Simpson, L. (1990) Guide RNAs in kinetoplastid mitochondria have a nonencoded 3' oligo(U) tail involved in recognition of the preedited region. *Cell* **62** 391-397.

Boer, P.H. & Gray, M.W. (1986a) Nucleotide sequence of a protein coding region in *Chlamydomonas reinhardtii* mitochondrial DNA. *Nucleic Acids Res.* **14** 7506-7507.

Boer, P.H. & Gray, M.W. (1986b) The URF5 gene of *Chlamydomonas reinhardii* mitochondria: DNA sequence and mode of transcripiion. *EMBO J.* **5** 21-28.

Boer, P. H. & Gray, M. W. (1988) Genes encoding a subunit of respiratory NADH dehydrogenase (ND 1) and a reverse transcriptase-like protein (RTL) are linked to ribosomal RNA gene pieces in *Chlamydomonas reinhardtii* mitochondrial DNA. *EMBO J.* **7** 3501-3508.

Boer, P.H., McIntosh, J.E., Gray, M.W. & Bonen, L. (1985) The wheat mitochondrial gene for apocytochrome b: absence of a prokaryotic ribosome binding site. *Nucleic Acids Res.* **13** 2281-2292.

Bonen, L. & Bird, S. (1988) Sequence analysis of the wheat mitochondrial atp6 gene reveals a fused upstream reading frame and markedly divergent N-termini among plant ATP6 proteins. *Gene* **73** 47-56.

Bonen, L., Bird, S. & Belanger, L. (1990) Characterization of the wheat mitochondrial orf25 gene. *Plant Mol. Biol.* **15** 793-795.

Bonen, L., Boer, P.H. & Gray, M.W. (1984) The wheat cytochrome oxidase subunit II gene has an intron insert and three radical aminoacid changes relative to maize. *EMBO J.* **3** 2531-2536.

Bonnard, G., Gualberto, J.M., Lamattina, L. & Grienenberger, J.M. (1992) RNA editing in plant mitochondria. *CRC Crit. Rev. Plant Sci.* **10** 503-524.

Cech, T. R. (1991) RNA editing: world's smallest introns? *Cell* **64** 667-669.

Chapdelaine Y. & Bonen, L. (1991) The wheat mitochondrial gene for subunit-I of the NADH dehydrogenase complex - a trans-splicing model for this gene-in-pieces. *Cell* **65** 465-472.

Chen, S.I.H., Habib, G., Yang, C.Y., Gu, Z.W., Lee, B.R., Weng, S.A., Silbennan, S.R., Cai, S.J., Deslypere, J.P., Rosseneu, M., Gotto, A.M., Li, W.H. & Chan, L. (1987) Apolipoprotein B-48 is the product of a messenger RNA with an organspecific in-frame stop codon. *Science* **238** 363-366.

Conklin, P.L. & Hanson, M.R. (1991) Ribosomal protein-S19 is encoded by the mitochondrial genome in *Petunia hybrida*. *Nucleic Acids Res* **19** 2701-2705.

Conklin, P.L., Wilson, R.K. & Hanson, M.R. (1991) Multiple trans-splicing events are required to produce a mature nad1 transcript in a plant mitochondrion. *Gene Develop.* **5** 1407-1415.

Covello, P.S. & Gray, M.W. (1989) RNA editing in plant mitochondria. *Nature* **341** 662-666.

Covello, P.S. & Gray, M.W. (1990a) Differences in editing at homologous sites in messenger RNAs from angiosperm mitochondria. *Nucleic Acids Res.* **18** 5189-5196.

Covello, P.S. & Gray, M.W. (1990b) RNA sequence and the nature of the Cua binding site in cytochrome c oxidase. *FEBS Lett.* **268** 5-7.

Desouza, A.P., Jubier, M.F., Delcher, E., Lancelin, D. & Lejeune, B. (1991) A transsplicing model for the expression of the tripartite nad5 gene in wheat and maize mitochondria. *Plant Cell* **3** 1363-1378.

Dewey, R.E., Levings III, C.S. & Timothy, D.H. (1986) Novel recombinations in the maize mitochondrial genome produce a unique transcriptional unit in the Texas male sterile cytoplasm. *Cell* **44** 439-449.

Douce, R. & Neuburger, M. (1989) The uniqueness of plant mitochondria. *Annu. Rev. Plant Physiol. Plant Mol. Biol.* **40** 371-414.

Driscoll, D.M., Wynne, J.K., Wallis, S.C. & Scott, J. (1989) An *in vitro* system for the editing of apolipoprotein B mRNA. *Cell* **58** 519-525.

Fox, T.D. & Leaver, C.J. (1981) The *Zea mays* mitochondrial gene coding cytochrome oxidase subunit II has an intervening sequence and does not contain TGA codons. *Cell* **26** 315-323.

Gay, N.J. & Walker, J.M. (1985) Two genes encoding the bovine mitochondrial ATP synthase proteolipid specify precursors with different import sequences and are expressed in a tissue specific manner. *EMBO J.* **4** 3519-3524.

Glaubitz, J.C. & Carlson, J.E. (1992) RNA editing in the mitochondria of a conifer. *Curr. Genet.* **22** 163-165.

Graves, P.V., Bégu, D., Velours, J., Neau, E., Belloc, F., Litvak, S. & Araya, A. (1990) Direct protein sequencing of wheat mitochondrial ATP synthase subunit 9 confirms RNA editing in plants. *J. Mol. Biol.* **214** 1-6.

Gray, M.W. (1989) Origin and evolution of mitochondrial DNA. *Annu. Rev. Cell Biol.* **5** 25-50

Gray, M.W., Cedergren, R., Abel, Y. & Sankoff, D. (1989) On the evolutionary origin of the plant mitochondrion and its genome. *Proc. Natl. Acad. Sci. USA* **86** 2267-2271.

Gray, M.W., Hanic-Joyce, P.J. & Covello, P.S. (1992) Transcription, processing and editing in plant mitochondria. *Annu. Rev. Plant Physiol. Plant Mol. Biol.* **43** 145-175.

Gruissem, W., Barkan, A., Deng, X.W. & Stern, D. (1988) Transcriptional and posttranscriptional control of plastid mRNA levels in higher plants. *Trends Genet.* **4** 258-263.

Gualberto, J.M., Wintz, H., Weil, J.H. & Grienenberger, J.M. (1988) The genes coding for subunit 3 of NADH dehydrogenase and for ribosomal protein S12 are present in the wheat and maize mitochondrial genomes and are co-transcribed. *Mol. Gen. Genet.* 215 118-127.

Gualberto, J.M., Lamattina, L., Bonnard, G., Weil, J.H. & Grienenberger, J.M. (1989) RNA editing in wheat mitochondria results in the conservation of protein sequences. *Nature* **341** 660-662.

Gualberto, J.M., Domon, C., Weil, J.H. & Grienenberger, J.M. (1990a) Structure and transcription of the gene coding for subunit 3 of cytochrome oxidase in wheat mitochondria. *Curr. Genet.* **17** 41-47.

Gualberto, J.M., Weil, J.H. & Grienenberger, J.M. (1990b) Editing of the wheat coxlll transcript: evidence for twelve C-to-U and one U-to-C conversions and for sequence similarities around editing sites. *Nucleic Acids Res.* **18** 3771-3776.

Gualberto J.M. Weil, J.H. & Grienenberger, J.M. (1990c) Structure and expression of a wheat mitochondrial transcription unit. In Structure, function and biogenesis of energy transfer systems, Quagliariello, E., *et al.*, eds. Elsevier Science Publishers, pp. 113-116.

Gualberto, J.M., Bonnard, G., Lamattina, L. & Grienenberger, J.M. (1991) Expression of the wheat mitochondrial nad3-rpsl2 transcription unit: correlation between editing and mRNA maturation. *Plant Cell* **3** 1109-1120.

Gwynn, B., Dewey, R.E., Sederoff, R.R., Timothy, D.H. & Levings III, C.S. (1987) Sequence of the 18S-5S ribosomal gene region and the cytochrome oxidase II gene from mtDNA of *Zea diploperennis, Theor. Appl. Genet.* **74** 781-788.

Handa, H. & Nakajima, K. (1992) Different organization and altered transcription of the mitochondrial atp6 gene in the male-sterile cytoplasm of rapeseed (*Brassica-Napus L*). *Curr. Genet.* **21** 153-159.

Hanic-Joyce, P.J. & Gray, M.W. (1990) Processing of transfer RNA precursors in wheat mitochondrial extract. *J. Biol. Chem.* **265** 13782-13791.

Hanic-Joyce, P.J. & Gray, M.W. (1991) Accurate transcription of a plant mitochondrial gene *in vitro. Mol. Cell. Biol.* **11** 2035-2039.

Hanson, M.R. (1991) Plant mitochondrial mutations and male sterility. *Annu. Rev. Genet.* **25** 461-486.

Haouazine, N., Pereira de Souza, A., Jubier, M. F., Lancelin, D., Delcher, D. & Lejeune, B. (1992) The wheat mitochondrial genome contains an ORF showing sequence homology to the gene encoding the subunit 6 of the NADH-ubiquinone oxidoreductase. *Plant Mol. Biol.* **20** 395-404.

Hensgens, L.A., Grivell, L.A., Borst, P. & Bos, J.L. (1979) Nucleotide sequence of the mitochondrial structural gene for subunit 9 of yeast ATPase complex. *Proc. Natl. Acad. Sci. USA* **76** 1663-1667.

Hernould, M., Mouras, A., Litvak, S. & Araya, A. (1992) RNA editing of the mitochondrial atp9 transcript from tobacco. *Nucleic Acids Res.* **20** 1809.

Hiesel, R. & Brennicke, A. (1983) Cytochrome oxydase subunit II in mitochondria of *Oenothera* has no intron. *EMBO J.* **2** 2173-2178.

Hiesel, R. & Brennicke, A. (1987) cDNA cloning of mitochondrial transcripts from *Oenothera. Plant Science* **51** 225-230.

Hiesel, R., Wissinger, B., Schuster, W. & Brennicke, A. (1989) RNA editing in plant mitochondria. *Science* **246** 1632-1634.

Hiesel, R., Wissinger, B. & Brennicke, A. (1990) Cytochrome oxidase subunit II mRNAs in *Oenothera* mitochondria are edited at 24 sites. *Curr. Genet.* **18** 371-375.

Hiratsuka, J., Shimada, H., Whittier, R., Ishibashi, T., Sakamoto, M., Mori, M., Kondo, C., Honji, Y., Sun, C.R., Meng, B.Y., Li, Y.Q., Kanno, A., Mshikawa, Y., Hirai, A., Shinozaki, K. & Sugiura, M. (1989) The complete sequence of the rice (*Oryza sativa*) chloroplast genome: intermolecular recombination between distinct tRNA genes accounts for a major plastid DNA inversion during the evolution of cereals. *Mol. Gen. Genet.* **217** 185-194.

Hoch, B., Maier, R.M., Appel, K., Igloi, G.L. & Kossel, H. (1991) Editing of a chloroplast mRNA by creation of an initiation codon. *Nature* **353** 178-180.

Hodges, P.E., Navaratnam, N., Greeve, J.C. & Scott, J. (1991) Site-specific creation of uridine from cytidine in apolipoprotein B mRNA editing. *Nucleic Acids Res.* **19** 1197-1201.

Hunt, M.D. & Newton, K.J. (1991) The NCS3 mutation: genetic evidence for the expression of ribosomal protein genes in Zeamays mitochondria. *EMBO J.* **10**

1045-1052.

Kao, T.H., Moon E. & Wu R. (1984) Cytochrome oxidase subunit II gene of rice has an insertion sequence with intron. *Nucleic Acids Res.* **12** 7305-7315.

Kempken, F., Mullen, J.A., Pring, D.R. & Tang, H.V. (1991) RNA editing of Sorghum mitochondrial atp6 transcripts changes 15 amino acids and generates a carboxy-terminus identical to yeast. *Curr. Genet.* **20** 417-422.

Knoop, V., Schuster, W., Wissinger, B. & Brennicke, A. (1991) Trans-splicing integrates an exon of 22 nucleotides into the nad5 messenger RNA in higher plant mitochondria. *EMBO J.* **10** 3483-3493.

Kudla, J., Igloi, G., Metzlaff, M., Hagemann, R. & Kossel, H. (1992) RNA editing in tobacco chloroplasts leads to the formation of a translatable psbL messenger RNA by a C-to-U substitution within the initiation codon. *EMBO J.* **11** 1099-1103.

Kuntz, M., Camara, B., Weil, J.H. & Schantz, R. (1992) The psbL gene from bell pepper (*Capsicum annuum*) plastid RNA editing also occurs in nonphotosynthetic chromoplasts. *Plant Mol. Biol.* **20** 1185-1188.

Lamattina, L. & Grienenberger, J.M. (1991) RNA editing of the transcript coding for subunit-4 of NADH dehydrogenase in wheat mitochondria - uneven distribution of the editing sites among the 4 exons. *Nucleic Acids Res.* **19** 3275-3282.

Lamattina, L., Weil, J.H. & Grienenberger, J.M. (1989) RNA editing at a splicing site of NADH dehydrogenase subunit IV gene transcript in wheat mitochondria. *FEBS Lett.* **258** 79-83.

Lonsdale, D.M. (1989) The plant mitochondrial genome, in The biochemistry of plants: a comprehensive treatise, Stump, P. K. & Conn, E.E., eds. Academic Press, pp. 229-295.

Lonsdale, D.M. & Leaver, C.J. (1988) Mitochondrial gene nomenclature. *Plant Mol. Biol. Rep.* **6** 14-21.

Maier, R.M., Hoch, B., Zeltz, P. & Kossel, H. (1992) Internal editing of the maize chloroplast ndhA transcript restores codons for conserved amino acids. *Plant Cell* **4** 609-616.

Maréchal, L., Guillemaut, P., Grienenberger, J.M., Jeannin, G. & Weil, J.H. (1985) Sequence and codon recognition of bean mitochondria and chloroplast tRNAs[Trp]: evidence for a high degree of homology. *Nucleic Acids Res.* **13** 4411-4416.

Maréchal, L., Runeberg-Roos, P., Grienenberger, J.M., Colin, J., Weil, J.H., Lejeune, B., Quétier, F. & Lonsdale, D.M. (1987) Homology in the region containing a tRNA[Trp] gene and a (complete or partial) tRNA[Pro] gene in wheat mitochondrial and chloroplast genomes. *Curr. Genet.* **12** 91-98.

Michel, F., Umesono, K. & Oseki, H. (1989) Comparative and functional anatomy of group II catalytic introns - a review. *Gene* **82** 5-30.

Mulligan, R.M. (1991) RNA editing - when transcript sequences change. *Plant Cell* **3** 327-330.

Mulligan, R.M., Léon, P. & Walbot, V. (1991) Transcriptional and posttranscriptional regulation of maize mitochondrial gene expression. *Mol.*

Cell. Biol. **11** 533-543.

Newton, K.J. (1988) Plant mitochondrial genomes: organization expression and variation. *Annu. Rev. Plant Physiol. Plant Mol. Biol.* **39** 503-532.

Nowak, C. & Kück, U. (1990) RNA editing of the mitochondrial atp9 transcript from wheat. *Nucleic Acids Res.* **18** 7164-7164.

Nugent, J.M. & Palmer, J.D. (1991) RNA-mediated transfer of the gene coxII from the mitochondrion to the nucleus during flowering plant evolution. *Cell* **66** 473-481.

Oda, K., Yamato, K., Ohta, E., Nakamura, Y., Takemura, M., Nozato, N., Akashi, K., Kanegae, T., Ogura, Y., Kohchi, T. & Ohyama, K. (1992) Gene organization deduced from the complete sequence of liverwort *Marchantia polymorpha* mitochondrial DNA - a primitive form of plant mitochondrial genome. *J. Mol. Biol.* **223** 1-7.

O'Donovan, G. & Neuhard, J. (1970) Pyrimidine metabolism in microorganisms. *Bacteriol. Rev.* **34** 278-343.

Ohyama, K., Fukuzawa, H., Kohchi, T., Shirai, H., Sano, T., Sano, S., Umesono, K., Shiki, Y., Takeuchi, M., Chang, Z., Aota, S., Inokuchi, H. & Ozeki, H. (1986) Chloroplast gene organization deduced from complete sequence of liverwort *Marchantia polymorpha* chloroplast DNA. *Nature* **322** 572-574.

Ohyama, K., Fukuzawa, H., Kohchi, T., Sano, T., Sano, S., Shirai, H., Umesono, K., Shiki, Y., Takeuchi, M., Chang, Z., Aota, S., Inokuchi, H. & Ozeki, H. (1988) Structure and organization of *Marchantia polymorpha* chloroplast genome I. Cloning and gene identification. *J. Mol. Biol.* **203** 281-298.

Ohyama, K., Ogura, Y., Oda, K., Yamato, K., Ohta, E., Nakamura, Y., Takemura, M., Nozato, N., Akashi, K., Kanegae, T. & Yamada, Y. (1991) Evolution of organellar genomes, in Evolution of life. Fossils, molecules, and culture, Osawa, S. & Honjo, T., eds., Springer-Verlag, pp. 187-198.

Okada, M. & Nishimura, S. (1979) Isolation and characterization of a guanine insertion enzyme, a specific tRNA transglycosylase from *Escherichia coli, J. Biol. Chem.* **254** 3061-3066.

Olsen, L.C., Aasland, R., Wittner, C.U., Krokan, H.E. & Helland, D.E. (1989) Molecular cloning of human uracil-DNA glycosylase, a highly conserved DNA repair enzyme. *EMBO J.* **8** 3121-3125.

Palmer, J.D. (1990) Contrasting modes and tempos of genome evolution in land plant organelles. *Trends Genet.* **6** 115-120.

Palmer, J.D. & Herbon, L.A. (1988) Plant mitochondrial DNA evolves rapidly in structure but slowly in sequence. *J. Mol. Evol.* **28** 87-97.

Price, J.V., Kieft, G.L., Kent, J.R., Sievers, E.L. & Cech, T.R. (1985) Sequence requirements for self-splicing of the *Tetrahymena thermophila* pre-ribosomal RNA. *Nucleic Acids Res.* **13** 1871-1889.

Pruitt, K.D. & Hanson, M.R. (1989) Cytochrome oxidase subunit II sequences in petunia mitochondria: two intron-containing genes and an intron-less pseudogene associated with cytoplasmic male sterility. *Curr. Genet.* **16** 281-292.

Rapp, W.D. & Stern, D.B. (1992) A conserved 11 nucleotide sequence contains an essential promoter element of the maize mitochondrial atp1 gene. *EMBO J.* **11** 1065-1073.

Salazar, R.A., Pring, D.R. & Kempken, F. (1991) Editing of mitochondrial atp9 transcripts from two Sorghum lines. *Curr. Genet.* **20** 483-486.

Sangaré, A., Weil, J.H. & Grienenberger, J.M. (1990) Maize and wheat mitochondrial tRNA^{Pro} coding regions have similar sequences but different organizations. *Biochim. Biophys. Acta* **1049** 96-98.

Schuster, W. & Brennicke, A., (1990) RNA editing of ATPase subunit 9 transcripts in *Oenothera* mitochondria. *FEBS Lett.* **268** 252-256.

Schuster, W. & Brennicke, A. (1991a) RNA editing in ATPase subunit-6 messenger RNAs in oenothera mitochondria - a new termination codon shortens the reading frame by 35 amino acids. *FEBS Lett.* **295** 97 -101.

Schuster, W. & Brennicke, A. (1991b) RNA editing makes mistakes in plant mitochondria - editing loses sense in transcripts of a rpsl9 pseudogene and in creating stop codons in coxl and rps3 messenger RNAs of *Oenothera*. *Nucleic Acids Res.* **19** 6923-6928.

Schuster, W., Hiesel, R., Isaac, P.G., Leaver, C.J. & Brennicke, A. (1986) Transcript termini of messenger RNAs in higher plant mitochondria. *Nucleic Acids Res.* **14** 5943-5954.

Schuster, W., Hiesel, R., Wissinger, B. & Brennicke, A. (1990a) RNA editing in the cytochrome b locus of the higher plant *Oenothera berteriana* includes a U-to-C transition. *Mol. Cell. Biol.* **10** 2428-2431.

Schuster, W., Unseld, M., Wissinger, B. & Brennicke, A. (1990b) Ribosomal protein S14 transcripts are edited in *Oenothera* mitochondria. *Nucleic Acids Res.* **18** 229-233.

Schuster, W., Wissinger, B., Unseld, M. & Brennicke, A. (1990c) Transcripts of the NADH-dehydrogenase subunit 3 gene are differentially edited in *Oenothera* mitochondria. *EMBO J.* **9** 263-269.

Schuster, W., Temes, R., Knoop, V., Hiesel, R., Wissinger, B. & Brennicke, A. (1991a) Distribution of RNA editing sites in *Oenothera* mitochondrial mRNAs and rRNAs. *Curr Genetics* **20** 397-404.

Schuster, W., Wissinger, B., Hiesel, R., Unseld, M., Gerold, E., Knoop, V., Marchfelder, A., Binder, S., Schobel, W., Scheike, R., Grönger, P., Ternes, R. & Brennicke, A. (1991b) Between DNA and protein - RNA editing in plant mitochondria. *Physiol. Plant.* **81** 437-445.

Shinozaki, K., Ohme, M., Tanaka, M., Wakasugi, T., Hayashida, N., Matsubayashi, T., Zaita, N., Chunwongse, J., Obokata, J., Yamaguchi-Shinozaki, K., Ohto, C., Torazawa, K., Meng, B.Y., Sugita, M., Deno, H., Kamogashira, T., Yamada, K., Kusuda, J., Takaiwa, F., Kato, A., Tohdoh, N., Shimida, H. & Sugiura, M. (1986) The complete nucleotide sequence of the tobacco chloroplast genome: its gene organization and expression. *EMBO J.* **5** 2043-2049.

Simpson, L. & Shaw, J. (1989) RNA editing and the mitochondrial cryptogenes of kinetoplastid protozoa. *Cell* **57** 355-366.

Sommer, B., Kohler, M., Sprengel, R. & Seeburg, P. H. (1991) RNA editing in brain controls a determinant of ion flow in glutamate-gated channels. *Cell* **67** 11-19.

Spencer, D.F., Schnare, M.N. & Gray, M.W. (1984) Pronounced structural similarities between the small subunit ribosomal RNA genes of wheat mitochondria and *Escherichia coli. Proc. Natl. Acad. Sci. USA* **81** 493-497.

Sutton, C.A., Conklin, P.L., Pruitt, K.D. & Hanson, M.R. (1991) Editing of pre-mRNAs can occur before cis- and trans-splicing in Petunia mitochondria. *Mol. Cell. Biol.* **11** 4274-4277.

Taylor, J., Zeng, H.-Q., Fu, T.-B., Turner, J., Ryu, W.-S., Bayer, M., Netter, H. & Lazinski, D. (1992) 18th EMBO Annual Symposium on RNA, Abstract book pp. 69-70.

Van Grinsven, M.Q.J.M. & Kool, A.J. (1988) Plastid gene regulation during development: an intriguing complexity of mechanisms. *Plant Mol. Biol. Rep.* **6** 213-239.

Walbot, V. (1991) RNA editing fixes problems in plant mitochondrial transcripts. *Trends Genet.* **7** 37-39.

Ward, G.C. & Levings III, C.S. (1991) The protein-encoding gene T-urf13 is not edited in maize mitochondria. *Plant. Mol. Biol.* **17** 1083-1088.

Wintz, H. & Hanson, M.R. (1991) A termination codon is created by RNA editing in the petunia mitochondrial atp9 gene transcript. *Curr. Genet.* **19** 61-64.

Wissinger, B., Schuster, W. & Brennicke, A. (1990) Species-specific RNA editing patterns in the mitochondrial rps13 transcripts of *Oenothera* and *Daucus. Mol. Gen. Genet.* **224** 389-395.

Wissinger, B., Schuster, W. & Brennicke, A. (1991) Trans splicing in Oenothera mitochondria: nad1 mRNAs are edited in exon and trans-splicing group-II intron sequences. *Cell* **65** 473-482.

Wissinger, B., Brennicke, A., and Schuster, W. (1992) Regenerating good sense, RNA editing and trans-splicing in plant mitochondria. *Trends Genet.* **8** 322-328.

Yang, A.J. & Mulligan, R.M. (1991) RNA editing intermediates of cox2 transcripts in maize mitochondria. *Mol. Cell. Biol.* **11** 4278-4281.

Zalkin H. (1985) CTP synthetase. *Methods Enzymol.* **113** 282-287.

Zheng, H., Fu, T.B., Lazinski, D. & Taylor, J. (1992) Editing on the genomic DNA of human hepatitis delta virus. *J. Virol.* **66** 4693-4697.

8

DOUBLE-STRANDED RNA ADENOSINE DEAMINASE: A POTENTIAL AGENT FOR RNA EDITING?

Unkyu Kim and Kazuko Nishikura
The Wistar Institute of Anatomy and Biology, 3601 Spruce Street, Philadelphia,
Pennsylvania 19104, USA

A novel cellular activity that specifically unwinds double-stranded RNA (dsRNA) was first detected in eggs and early embryos from *Xenopus laevis* as a result of experiments designed to follow the fate of artificially introduced RNA duplexes in the cell (Rebagliati and Melton, 1987; Bass and Weintraub, 1987). The activity, originally called "dsRNA unwindase" or "denaturase", was subsequently also found in a wide range of mammalian somatic cells, and is assumed to carry out essential biological functions (Wagner and Nishikura, 1988; Wagner *et al.*, 1990). One striking characteristic of this dsRNA-unwinding activity is the unprecedented nature of its unwinding mechanism. During the unwinding reaction, many adenosine residues are deaminated and thus converted to inosines (Wagner *et al.*, 1989; Bass and Weintraub, 1988). Accordingly, the activity has been termed dsRNA unwinding/modifying activity (Wagner *et al.*, 1989; Bass and Weintraub, 1988). In this article, we will refer to the activity as dsRNA adenosine deaminase, a term which reflects its hitherto unique enzymatic property.

Although speculative at this stage, several hypotheses have been proposed with regard to the possible biological function of this activity. In light of its ability to change the base composition in the substrate RNA, one obvious possibility is that the dsRNA adenosine deaminase may play a role in an RNA editing process. In this chapter, first a brief review will be given on the characteristics of the dsRNA adenosine deaminase known at present. Then, the possibility of involvement of the enzyme in the RNA editing process will be discussed in detail, including specific cases in which the dsRNA adenosine deaminase has been proposed as an RNA editing factor.

8.1 dsRNA ADENOSINE DEAMINASE

8.1.1 dsRNA-unwinding activity

An activity capable of specifically dissociating RNA duplexes was originally discovered in *Xenopus* eggs and embryos (Rebagliati and Melton, 1987; Bass and Weintraub, 1987). An identical activity was subsequently discovered in mammalian cells (Wagner and Nishikura, 1988). This activity seems to involve (a) protein(s), since treatment with proteinase K abolishes the activity (Bass and Weintraub, 1987; Wagner and Nishikura, 1988). Neither ATP nor any other nucleoside phosphate nor divalent cations are required for the activity (Bass and Weintraub, 1987; Wagner and Nishikura, 1988). The activity is absolutely

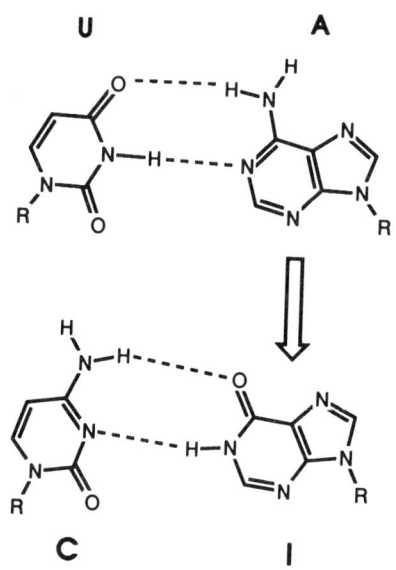

Fig. 1 — Watson-Crick basepairings of U-A and C-I. The conversion of adenosine to inosine results in a mismatched I-U base pair that must adopt a thermodynamically less stable, wobble hydrogen-bonding configuration. Inosine preferentially base pairs with cytosine in a Watson-Crick hydrogen-bonding configuration.

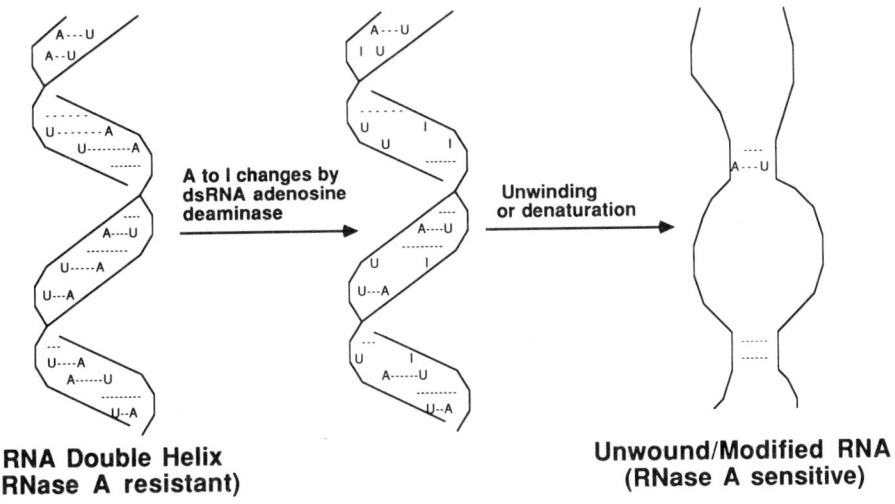

RNA Double Helix   Unwound/Modified RNA
(RNase A resistant)   (RNase A sensitive)

Fig. 2 —— Modification and unwinding of double-stranded RNA by the dsRNA adenosine deaminase. Unwinding of dsRNA through disruption of A-U base pairing by the unwindase is schematically shown. It is not known at present whether the modified and partly denatured reaction products are dissociated into two strands *in vivo*.

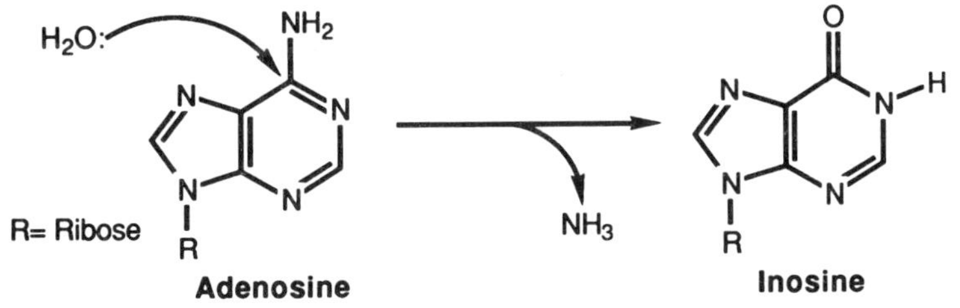

Fig. 3 —— Deamination of adenosine to inosine. An amino group at position 6 of the adenine ring is removed and replaced by a carbonyl group by the deamination process.

specific for, and appears to bind strongly to, dsRNA (Bass and Weintraub, 1987; Wagner and Nishikura, 1988). The activity recognizes the RNA double-helix structure without strict sequence specificity, as it is capable of unwinding different dsRNAs made from various sense and antisense RNAs synthesized *in vitro* containing ß-globin, c-myc, and chloramphenicol acetyl transferase sequences.

8.1.2 Deamination of adenosines in dsRNAs to inosines
The unwound RNA is incapable of rehybridizing back to dsRNA, indicating that the RNA is covalently altered during the unwinding reaction (Bass and Weintraub, 1987; Wagner *et al.*, 1989; Bass and Weintraub, 1988). The nature of the alteration in the unwound RNA was determined by analyzing the unwound and then strand-separated RNAs for modified bases. Thin-layer chromatography, as well as high-pressure liquid chromatography, indicated that a large fraction of adenosine residues of the unwound RNA is converted to inosine (Wagner *et al.*, 1989; Bass and Weintraub, 1988). Adenosines in both sense and antisense strands of unwound RNA are converted to inosines (Wagner *et al.*, 1989; Bass and Weintraub, 1988). Inosines can form a stable Watson-Crick base pair with hydrogen bonds to cytosines, whereas base pairing of inosines with uracils can be achieved only in a less stable wobble configuration (Fig. 1). Thus, it seems that accumulation of extensive mismatched I-U base pairs in the dsRNA may cause unwinding of the RNA double-helix (Fig. 2). This is the first and so far the only, known RNA-unwinding activity that results in an accompanying base modification on the substrate RNA. This dsRNA-unwinding/modifying activity further differs from other dsRNA-unwinding activities or RNA helicases in that it seems to bind specifically to dsRNA. Other known RNA helicases that contain conserved amino acid sequences termed DEAD boxes unwind dsRNAs by virtue of binding to single-stranded regions of the RNAs (Wassarman and Steitz, 1991).

   The biochemical mechanism underlying adenosine-to-inosine conversion seems to be deamination of an adenine base. Several enzymes involved in purine metabolism catalyze the conversion of adenine to hypoxanthine (the free base of inosine) by deamination at position 6 of the adenine ring (Fig. 3). These enzymes, including adenosine deaminase and 5'-adenylic acid deaminase, have rigid substrate specificity and were found to be incapable of modifying adenosines in dsRNA; conversely, the dsRNA-unwinding/modifying activity was found not to modify adenosines or adenylic acid (Bass and Weintraub, 1988; Wagner and Nishikura, unpublished results). Another known enzyme which introduces inosines in RNA is transfer RNA (tRNA) hypoxanthine ribosyltransferase. The enzyme posttranscriptionally substitutes a hypoxanthine for the pre-existing base, including adenosine, in the first position of an anti-codon of several tRNAs (Elliott and Trewyn, 1984). However, the dsRNA-unwinding/modifying activity differs from this type of enzyme, since it was shown that adenosine residues labeled with ATP carrying a ^{14}C at position 8 of the adenine ring remained labeled after conversion to inosine (Wagner *et al.*, 1989; Polson *et al.*, 1991). This suggests that no replacement of base takes place during the modifying reaction. Thus, it appears that the dsRNA-unwinding/modifying activity is indeed a novel adenosine deaminase specific for dsRNA, and hence the term "dsRNA adenosine deaminase."

8.1.3 Requirement for substrate dsRNAs

Although there is no strict sequence requirement for the substrate dsRNAs, some base preference for the 5' neighbor of the modified adenosine has been reported. When the unwound/modified basic fibroblast growth factor (bFGF) messenger RNA (mRNA) was examined *in vivo* after polymerase chain reaction (PCR) amplification and cDNA cloning, it was found that adenosines having adenosines or uracils as 5' neighbors were modified more frequently than those having guanosines or cytosines as neighbors (Kimelman and Kirschner, 1989). Thus, within a given dsRNA, adenosine residues in a stretch of consecutive As or an AT-rich region might be preferentially modified by the dsRNA adenosine deaminase.

Investigations on substrate specificity for the dsRNA adenosine deaminase have been conducted *in vitro* using an assay system and artificially synthesized RNA substrates. Intermolecular dsRNAs having blunt ends, 5' overhangs, and 3' overhangs, as well as partial dsRNAs, were modified equally well (Wagner *et al.*, 1989; Bass and Weintraub, 1988). The activity also did not distinguish between capped and non-capped RNAs (Wagner *et al.*, 1989; Bass and Weintraub, 1988). Both intermolecular and intramolecular dsRNAs serve as substrates for the dsRNA adenosine deaminase (Nishikura *et al.*, 1991). The dsRNA adenosine deaminase requires a double-stranded region of at least 15-20 base pairs for substrate recognition (Nishikura *et al.*, 1991). Finally, modification efficiency was found to be critically dependent on the length of the double-stranded region: when the size decreased below 100 base pairs, it dropped precipitously (Nishikura *et al.*, 1991). Efficient modification occurs, at least *in vitro*, only with relatively long dsRNA (>100 base pairs), perhaps because multiple copies of the enzyme must be bound (Nishikura *et al.*, 1991). Thus, naturally occurring viral and eukaryotic gene transcripts, which usually contain relatively short and often imperfect double-stranded regions, may not be efficient substrates. However, the naturally occurring hairpins in these RNAs may be stabilized or, in some way, altered *in vivo* to form more helix-like structures, and thus allow the dsRNA adenosine deaminase to recognize them as substrate RNAs (Nishikura *et al.*, 1991). One additional note of caution in interpreting the requirement *in vitro* is that the modification efficiency was determined by measuring how many adenosines were converted to inosines. It is possible that the function of the deaminase *in vivo* is to modify not many but only a few selected adenosines. The relevance of this point in considering the activity as a potential participant in RNA editing will be further elaborated in later sections.

8.2 CELLULAR AND TISSUE DISTRIBUTION OF THE ENZYME

8.2.1 Ubiquitous expression

Based on its ubiquitous presence in a wide range of cell types and species, the dsRNA adenosine deaminase seems to be a typical housekeeping protein, providing a fundamental function (Wagner *et al.*, 1990). A variety of primary tissues and transformed cell lines derived from different tissues were surveyed for the level of expression of the activity (Wagner *et al.*, 1990). Rat liver, kidney, spleen, testis, and lymph node tissues, as well as human natural killer cells and T cells, both isolated from peripheral blood, were all found to contain the activity (Wagner *et al.*, 1990). In addition to primary tissues and cells, a variety of transformed cell lines commonly used, including Cos, HeLa, and 3T3, were also analyzed, and were all found to contain the activity. Quiescent mouse fibroblast 3T3 cells increased their dsRNA adenosine deaminase activity three-fold after stimulation with serum. In addition, the level of the activity decreased four-fold *in vitro* during induction of the mouse muscle cell line C_2C_{12} from myoblast to myotube, although the activity was still detectable in the differentiated postmitotic myotubes (Wagner *et al.*, 1990).

The involvement of dsRNA adenosine deaminase has been suggested in a highly biased hypermutation of measles virus genomes that causes some diseases of the human central nervous system (CNS) (see below) (Bass *et al.*, 1989; Cattaneo *et al.*, 1988, 1989). Interestingly, extracts of human, calf, rat, or mouse brain contained high levels of the dsRNA adenosine deaminase, as did bovine oligodendrocytes and rat astrocytes. In addition to these primary tissues, various transformed cell lines derived from the central and peripheral nervous systems contained high levels of the activity. It was concluded that the dsRNA adenosine deaminase is detectable in all cells and tissues, including those of the CNS, regardless of the differentiated, mitotic, or transformed state of the cell (Wagner *et al.*, 1990).

8.2.2 Nuclear localization

In mammalian somatic cells, the dsRNA adenosine deaminase is localized exclusively in the nucleus (Wagner *et al.*, 1990). Interestingly, the activity in *Xenopus laevis*, which has been reported to reside in the nucleus of oocytes, is released into the cytoplasm during meiotic maturation and remains in the cytoplasm during fertilization and early embryogenesis (B. Bass, personal communication). It is currently not known exactly when the dsRNA adenosine deaminase returns to the nucleus of somatic cells during *Xenopus* embryogenesis. The prolonged presence of the activity in the cytoplasm of the *Xenopus* oocyte allows the dsRNA adenosine deaminase to participate in regulation of a particular maternal transcript. In maturing *Xenopus laevis* oocytes, the dsRNA formed between sense and antisense bFGF RNAs is modified in the cytoplasm by the dsRNA adenosine deaminase released from the nucleus after the breakdown of the germinal vesicle that accompanies meiotic maturation (Kimelman and Kirschner, 1989). It was proposed that the extensive modification (up to 50% of

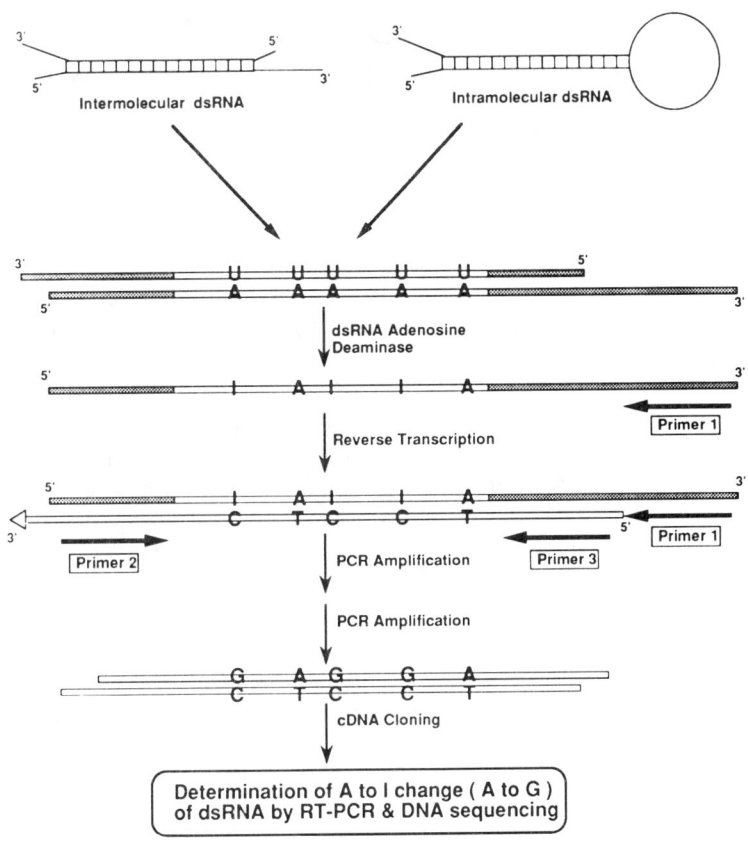

Fig. 4 — Experimental design for detection of A-to-I conversions *in vivo* in unwound/modified dsRNA. Intermolecular or intramolecular dsRNAs are unwound/modified *in vivo* by dsRNA adenosine deaminase. RNAs are converted to cDNA by reverse transcriptase, and then amplified by PCR. The resulting products are then cloned and sequenced. Note that inosines in the unwound/modified dsRNA are transcribed to guanosines in the final DNA products. All primers must be designed to hybridize to the region outside (hatched box) the double-stranded region (open box).

adenosines) is responsible for rapid degradation of the bFGF dsRNA (Kimelman and Kirschner, 1989). At present, it is unknown whether this type of modification/degradation process by the dsRNA adenosine deaminase occurs widely. Although it is likely that in somatic cells the activity has a greater chance to interact with nuclear dsRNAs, one cannot rule out the possibility that the activity may be released into the cytoplasm of somatic cells during mitosis and, thus, target cytoplasmic dsRNAs as in the case of bFGF RNAs in *Xenopus* oocytes.

8.3 INVOLVEMENT IN RNA EDITING

Currently, the physiological function of the dsRNA adenosine deaminase is not known. Although the activity may play a similar role as that in *Xenopus* oocytes, i.e. triggering RNA degradation in somatic cells, so far there has not been any observation made to support this hypothesis. Another possible role the activity may play is to participate in an RNA editing process. In this section, two cases of RNA editing in which the dsRNA adenosine deaminase might be involved are presented; see also Table 1. One case occurs in an RNA viral genome, another in rat brain cells. In both cases, the edited base is an A which is converted to a G. These mutations could be introduced by the dsRNA adenosine deaminase, since the A→I change is translated into an A→G change of the cDNA by reverse transcriptase (see Fig. 4). With replicating RNA viruses, an A→I change on the minus (genomic) strand will be perpetuated as a U→C change on the plus (anti-genomic) strand or mRNA, and ultimately, as an A→G change on the minus strand following the replication of the plus strand (see Bass *et al.*, 1989). As discussed earlier, the dsRNA adenosine deaminase requires double-stranded RNA as the substrate. The origin of dsRNA is considered for both cases.

8.3.1 Biased hypermutation of RNA viruses

It has been suggested that several biased hypermutations in RNA viral genomes are caused by the dsRNA adenosine deaminase (Bass *et al.*, 1989; Cattaneo *et al.*, 1989). Subacute sclerosing panencephalitis and measles inclusion-body encephalitis are two rare forms of persistent infection of the human CNS by the measles virus. Both infections develop years after acute infection and finally result in progressive cerebral degeneration, leading to death (Cattaneo *et al.*, 1988, 1989). Antibody-staining studies led to the supposition that defective production of several proteins involved in viral budding may cause the persistent infection (Cattaneo *et al.*, 1988). Recently, by cDNA cloning and sequencing of the viral RNAs, it was found that extensive base changes had occurred in viral genomes isolated from diseased human brains (Cattaneo *et al.*, 1988, 1989). One of the most striking base transitions occurred in the case of measles inclusion-body encephalitis. In the genome (minus strand) coding for the matrix protein responsible for viral assembly, 50% of A residues (132 of 266 As) were mutated to G (Cattaneo *et al.*, 1988). In the case of subacute sclerosing panencephalitis, viral genes coding for two envelope glycoproteins, the fusion and hemagglutination proteins, were found to be mutated (Cattaneo *et al.*, 1989). A cluster of A→G mutations (20 of 97 As) were noted in the hemagglutination gene. One more case of A→G mutations was detected in a defective interfering particle of another RNA virus, vesicular stomatitis virus (O'Hara *et al.*, 1984). In the RNA viral genome, the minus strand is transcribed into mRNA, and also serves as a template for replication into the plus strand, which is then replicated into the minus strand. Modification may occur during the transcription or

Table 1 — RNA editing process in which the dsRNA adenosine deaminase may participate

Example	Description	Reference
Biased hypermutation in RNA virus genome.	Extensive A to G mutation: defective expression of viral genes.	O'Hara et al. (1984) Bass et al. (1989) Cattaneo et al. (1989) Cattaneo et al. (1988)
RNA editing of transcripts of glutamate-gated ion channel subunit gene.	Replacement of glutamine codon (CAG) by arginine codon (CGG).	Sommer et al. (1991)

replication of the RNA genome by "accidental" formation of dsRNAs (either minus strand-nascent transcript or minus strand-plus strand pairs) (see Bass et al., 1989). In addition, in the case of the defective interfering particle of vesicular stomatitis virus, various intramolecular dsRNA structures are predicted to form (O'Hara et al., 1984).

The extensive mutation drastically changed the coding capability of the matrix and hemagglutinin gene and resulted in the defective expression of these genes, including the alteration of the protein initiation codon in some cases (Cattaneo et al., 1988; O'Hara et al., 1984). Thus these cases may not fit into RNA editing in a strict sense. The lack of the functional viral envelope protein may help the virus to escape immune-system recognition and thus allow persistent infection (Cattaneo et al., 1988). It may be that biased hypermutation caused by the dsRNA adenosine deaminase occurs relatively frequently, but is propagated and detected only in the genome of persistent viruses or defective interfering particles because of the lower selective pressures operating in these infections. The frequency and extent of biased hypermutation detected may be related to the dsRNA adenosine deaminase level present in different tissues and the life cycle of individual RNA viruses (Cattaneo et al., 1989).

8.3.2 RNA editing of glutamate-gated ion channel gene transcripts

A family of ion-channel subtypes in the vertebrate nervous system responds to L-glutamate, a major neurotransmitter, and mediates fast excitatory synaptic responses (Sommer et al., 1991). The physiological functions of these glutamate-gated ion channels include the establishment and maintenance of synaptic plasticity underlying learning and memory, and, also, mediation of cell death following excessive glutamate release in the CNS in acute pathological situations such as trauma or stroke (Sommer et al., 1991). So far, six closely related subunits have been discovered in rat brain (named GluR-A, -B, -C, -D, and GluR-5, -6) (Sommer et al., 1991). All subunits contain four putative membrane-spanning regions (TMI-TMIV) which determine the characteristics of

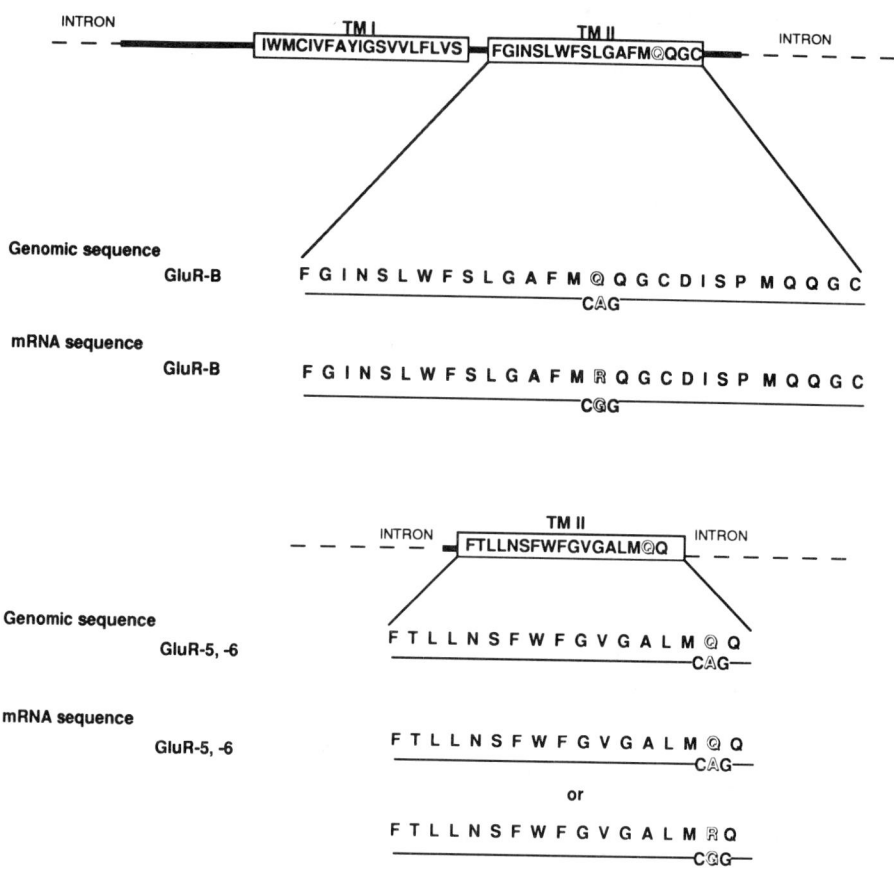

Fig. 5 — RNA editing of glutamate-gated ion-channel subunit transcripts. Comparison of genomic and cDNA sequences in the transmembrane domain II (TMII) region of the subunits GluR-B, -5, and -6. Capital letters in the box are single-letter codes for amino acids. Q (glutamine) in genomic sequences is converted to R (arginine) in cDNA sequences by an A-to-G change in the nucleotide sequence of these codons. (Modified from Fig. 2 of Sommer *et al.*, 1991).

the channel. The sequences of these transmembrane regions are well conserved among different subunits (Sommer *et al.*, 1991; Verdoorn *et al.*, 1991). Functional studies using recombinant subunits indicated that GluR-B exhibits current-voltage relationships distinct from GluR-A, -C, and -D (Verdoorn *et al.*, 1991). The difference is attributed to the presence of a positively charged amino acid (arginine) in the putative transmembrane domain TMII of GluR-B subunit. Other subunits studied, GluR-A, -C, and -D, had a neutral glutamine in place of arginine. Except for the arginine, the sequences of the TMII region in all subunits are identical (Verdoorn *et al.*, 1991). Site-directed mutagenesis showed that a mutant GluR-D subunit containing the arginine exhibits current-voltage relationships identical to those of Glu-R-B, thus confirming the arginine as the determining factor for the glutamate-gated channel behavior (Verdoorn *et al.*, 1991). In addition to GluR-B, a subpopulation of GluR-5 and -6 subunits also contain arginine in the TMII region (Sommer *et al.*, 1991).

A recent report suggests that the arginine in the TMII region of the glutamate-gated channel subunits is generated by an RNA editing process (Sommer *et al.*, 1991). A single nucleotide mismatch between genomic DNA and cDNA was noticed when murine genomic clones of GluR-B, -5, and -6 were sequenced. In place of CGG, which encodes arginine, it was found that genomic sequences contained CAG, a glutamine codon (Sommer *et al.*, 1991) (Fig. 5). The possibility of multiple genes or multiple exons was ruled out on the basis of Southern analysis of genomic DNA, and by the fact that no arginine codon containing genomic clone was found even though multiple copies of overlapping clones were analyzed (Sommer *et al.*, 1991). Thus it seems that genes for GluR-B, -5, and -6 all code only for glutamine at the site in TMII, and that the arginine codon in the mRNA is generated by conversion of the glutamine codon, CAG, to an arginine codon, CGG, posttranscriptionally. This is the first and so far only known case in which an A-to-G change is observed in mammalian transcripts. This A-to-G conversion detected by cDNA cloning and sequencing might be mediated by deamination of adenosine to inosine (see Fig. 4), and, if so, the dsRNA adenosine deaminase may be a candidate key factor. In this case, the resulting CIG in the mRNA may encode an arginine residue since an inosine is recognized as a guanosine by the translation factors (Basillo *et al.*, 1962). In order to establish the involvement of the dsRNA adenosine deaminase as the editing factor for GluR-B, -5, and -6 transcripts, one must demonstrate formation of dsRNA within the edited site. It is unknown at present whether an antisense RNA or a guide RNA complementary to the edited site is made in the cells. However, the conservation of sequences in TMII region among different subunits makes it difficult to explain why only the GluR-B, -5, and -6 transcripts are modified (Sommer *et al.*, 1991). Thus there could be variation in intron sequences, which could result in formation of the intramolecular dsRNA structures in some GluR subtype transcripts but not in others (Sommer *et al.*, 1991). Alternatively, the selectivity could be achieved by differential expression or availability of the dsRNA adenosine deaminase in different cell types of brain, since it has been shown by *in situ* hybridization that GluR subtype transcripts are differentially localized (Sommer *et al.*, 1991).

8.4 PERSPECTIVES

So far, only two cases of RNA editing in mammalian cellular RNA have been reported. One is the tissue-specific editing of human intestinal apolipoprotein B mRNA in which a genomically encoded C residue is converted to U (Chen *et al.*, 1987; Powell *et al.*, 1987; see Chapter 6). As discussed in Chapter 6, characterization of this editing process indicates that deamination of cytosine to uracil may be involved (Chen *et al.*, 1987; Powell *et al.*, 1987; Greeve *et al.*,

1991). The second case is the editing of glutamate-gated ion channel RNAs. In this case, a specific A residue is converted to G (Sommer *et al.*, 1991). This conversion may also involve a deamination process, namely deamination of the adenosine generating an inosine which will be recognized as a guanosine by reverse transcriptase as well as translation factors. (The possible parallels in the activity of both deaminating enzymes have been discussed in Chapter 6.) If this mechanism is genuine, then the dsRNA adenosine deaminase is likely to play a role in the editing process. However, two points must be resolved in order to establish the dsRNA adenosine deaminase as a *bona fide* editing factor. First, any substrate RNA for the enzyme must form RNA-RNA duplexes. This could be achieved by either an antisense RNA complementary to the edited site, or an intramolecular dsRNA such as a hairpin secondary structure. Second, the question must be answered of how specificity, which is crucial for RNA editing, is achieved. This is, at first glance, a particularly difficult question, considering the apparent lack of sequence specificity of the enzyme *in vitro* (Wagner and Nishikura, 1988; Wagner *et al.*, 1990; Nishikura *et al.*, 1991) and *in vivo* (Rebagliati and Melton, 1987; Bass and Weintraub, 1987; Kimelman and Kirschner, 1989). However, it is possible that the specificity is obtained by the expression of specific antisense RNAs or the formation of specific intramolecular dsRNAs containing only the A residue that is to be edited. As discussed earlier, another way of achieving specificity is to control the availability of the edited site to the dsRNA adenosine deaminase. Certain cellular factors, as yet unidentified, may bind to the dsRNA and mask the sites from the enzyme. These factors may be regulated (possibly by interacting with the dsRNA adenosine deaminase) in such a way as to allow interaction between the dsRNA adenosine deaminase and specific residues in the RNA.

Although there has not been any direct evidence that the dsRNA adenosine deaminase is in fact an editing factor, all the circumstantial observations indicate that the possibility exists. In light of the ubiquitous presence of the enzyme in a wide range of mammalian somatic cells, one may expect to discover more cases of RNA editing involving adenosine-to-guanosine conversion in mammalian transcripts. We have recently isolated human cDNA and bovine genomic clones coding for the dsRNA adenosine deaminase (Kim and Nishikura, unpublished results). Molecular genetic approaches using these clones may allow us more directly to test the potential of the dsRNA adenosine deaminase for RNA editing in the near future.

ACKNOWLEDGEMENTS
We thank J. M. Murray and R. Ricciardi for helpful discussion and critical reading of this manuscript. We also thank the Editorial Services Department of The Wistar Institute for preparing the manuscript. This project was supported by grants CA09171, CA10815, and GM40536 from the National Institutes of Health.

REFERENCES
Basillo, C., Wahba, A. J., Lengyel, P., Speyer J. F. & Ochoa, S. (1962). Synthetic polynucleotides and the amino acid code, V. *Proc. Natl. Acad. Sci.*

USA **48**: 613-616.

Bass, B. L. & Weintraub, H. (1987). A developmentally regulated activity that unwinds RNA duplexes. *Cell* **48**: 607-613.

Bass, B. L. & Weintraub, H. (1988). An unwinding activity that covalently modifies its double-stranded RNA substrate. *Cell* **55**: 1089-1098.

Bass, B. L., Weintraub, H., Cattaneo, R. & Billeter, M. A. (1989). Biased hypermutation of viral RNA genomes could be due to unwinding/modification of double-stranded RNA. *Cell* **56**: 331.

Cattaneo, R., Schmid, A., Eschle, D., Baczko, K., Meallen V. & Billeter, M. A. (1988). Biased hypermutation and other genetic changes in defective measles viruses in human brain infections. *Cell* **55**: 255-265.

Cattaneo, R., Schmid, A., Spielhofer, P., Kaelin, K., Baczko, K., Meulen, V., Pardowitz, J., Flanagan, S., Rima, B. K., Udem S. A. & Billeter, M. A. (1989). Mutated and hypermutated genes of persistent measles viruses which caused lethal human brain diseases. *Virology* **173**: 415-425.

Chen, S.-H., Habib, G., Yang, C.-Y., Gu, Z.-W., Lee, B. R., Weng, S.-A., Silberman, S. R., Cai, S.-J., Deslypere, J. P., Rosseneu, M., Gotto, A. M. Jr., Li, W.-H. & Chan, L. (1987). Apolipoprotein B-48 is the product of a messenger RNA with an organ-specific in-frame translational stop codon. *Science* **328**: 363-366.

Elliott, M. S. & Trewyn, R. W. (1984). Inosine biosynthesis in transfer RNA by an enzymatic insertion of hypoxanthine. *J. Biol. Chem.* **259**: 2407-2410.

Greeve, J., Navartnam, N. & Scott, J. (1991). Characterization of the apolipoprotein B mRNA editing enzyme; no similarity to the proposed mechanism of RNA editing in kinetoplastid protozoa. *Nucleic Acids Res.* **19**: 3569-3576.

Kimelman, D. & Kirschner, M. W. (1989). An antisense mRNA directs the covalent modification of the transcript encoding fibroblast growth factor in *Xenopus* oocytes. *Cell* **59**: 687-696.

Nishikura, K., Yoo, C., Kim, U., Murray, J. M., Estes, P. A., Cash, F. E. & Liebhaber, S. A. (1991). Substrate specificity of the dsRNA-unwinding/modifying activity. *EMBO J.* **10**: 3523-3532.

O'Hara, P. J., Nichol, S. T., Horodyski, F. M. & Holland, J. J. (1984). Vesicular stomatitis virus defective interfering particles can contain extensive genomic sequence rearrangements and base substitutions. *Cell* **36**: 915-924.

Polson, A. G., Crain, P. F., Pomerantz, S. C., McCloskey, J. A. & Bass, B. L. (1991). The mechanism of adenosine to inosine conversion by the double-stranded RNA-unwinding/modifying activity: a high-performance liquid chromatography-mass spectrometry analysis. *Biochemistry* **30**, 11507-11514.

Powell, L. M., Wallis, S. C., Pease, R. J., Edwards, Y. H., Knott, T. J. & Scott, J. (1987). A novel form of tissue-specific RNA processing produces apolipoprotein-B48 in intestine. *Cell* **50**: 831-840.

Rebagliati, M. R. & Melton, D. A. (1987). Antisense RNA injections in fertilized frog eggs reveal an RNA duplex unwinding activity. *Cell* **48**: 599-605.

Sommer, B., Köhler, M., Sprengel, R. & Seeburg, P. H. (1991). RNA editing in brain controls a determinant of ion flow in glutamate-gated channels. *Cell* **67**: (11-19).

Verdoorn, T. A., Burnshev, N., Monyer, H., Seeburg, P. H. & Sakmann, B. (1991). Structural determinants of ion flow through recombinant glutamate receptor channels. *Science* **252**: 1715-1718.

Wagner, R. W. & Nishikura, K. (1988). Cell cycle expression of RNA duplex unwindase activity in mammalian cells. *Mol. Cell. Biol.* **8**: 770-777.

Wagner, R. W., Smith, J. E., Cooperman, B. S. & Nishikura, K. (1989). A double-stranded RNA-unwinding activity introduces structural alterations by means of adenosine to inosine conversions in mammalian cells and *Xenopus* eggs. *Proc. Natl. Acad. Sci. USA* **86**: 2647-2651.

Wagner, R. W., Yoo, C., Wrabetz, L., Kamholz, J., Buchhalter, J., Hassan, N. F., Khalili, K., Kim, S. U., Perussia, B., McMorris, F. A. & Nishikura, K. (1990). Double-stranded RNA-unwinding and modifying activity is detected ubiquitously in primary tissue and cell lines. *Mol. Cell. Biol.* **10**: 5586-5590.

Wassarman, D. A. & Steitz, J. A. (1991). Alive with DEAD proteins. *Nature* **349**: 463-464.

Index